CICS Basic Training for Application Developers

Using DB2 and VSAM

Robert Wingate

ISBN 13: 9781734584769

Disclaimer

The content of this book is based upon the author's understanding of and experience with the following IBM products: CICS, DB2 and VSAM. Every attempt has been made to provide correct information. However, the author and publisher do not guarantee the accuracy of every detail, nor do they assume responsibility for information included in or omitted from it. All of the information in this book should be used at your own risk.

Copyright

Table of Contents

Introduction

Congratulations on your purchase of **CICS Basic Training for Application Developers.**
This book will help you learn the essential information you need to become productive
with CICS as soon as possible. You'll receive instruction, examples and questions/answers
to help you learn and to gauge your readiness for development work on an IBM mainframe
CICS technical team.

Why CICS?

CICS is an older product developed in the 1960's. However, there is still a large installed
base of CICS applications that must be supported. So there is still a need for CICS
programmers.

Frankly, CICS is not particularly intuitive. It can be pretty cryptic until you get some
experience developing an application. In this book, we'll go through the basics and develop
a few programs to get you started. By the time we finish, you should be well on the way to
working productively with CICS applications.

What This Book is Not

This is not an "everything you'll ever need to know about CICS" book. This text will teach
you what you need to know to become **productive quickly** with CICS. For additional
detail, you can download and reference the IBM manuals and Redbooks associated with
CICS. Also see the Additional Resources section at the end of the book for more web
resources.

Assumptions:

While I do not expect that you know much about CICS, I do assume that you've worked
on an IBM mainframe and know your way around. Also I assume that you have a working
knowledge of the COBOL programming language and the DB2 RDBMS which we will
use for all the embedded SQL examples. All in all, I assume you have:

1. A working knowledge of ISPF and JCL.

2. A working knowledge of COBOL.

3. A basic understanding of DB2 SQL.

4. Access to a mainframe computer running z/OS including CICS, VSAM and DB2.

How to Use This Book

We'll give you the basics of CICS including programming examples. Of course read the book, but you really need to try to follow along and do some of the programming examples yourself. If your current job includes access to CICS, my suggestion is to use it to learn. You may have to convince your supervisor (or other powers that be) to give you access to CICS if your current role does not include it. Employers will ordinarily be happy to provide an opportunity for someone who shows the initiative to get training and become more competent.

If you do not have access to a mainframe system through your job, I can recommend Mathru Technologies. You can rent a mainframe account from them at a very affordable rate, and this includes access to COBOL, CICS, VSAM and DB2 (at this writing they offer **DB2 version 10**). The URL to the Mathru web site is:

http://mathrutech.com/index.html

You'll need 3270 emulation software to connect to and interact with the mainframe environment. For example you might acquire IBM's Personal Communications product. There are also free 3270 emulator products on the web.

To Your Success

Knowledge and experience. Will that guarantee that you'll succeed as a CICS application developer? Of course, nothing is guaranteed in life. But if you put sufficient effort into a well-rounded study plan that includes both of the above, I believe you have a very good chance of excelling in the CICS world as an application developer. This is your chance to get a quick start!

Best of luck!

Robert Wingate

IBM Certified Application Developer – DB2 11 for z/OS

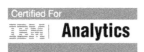

Chapter One: CICS Basics

Introduction
CICS is an acronym for Customer Information Control System. This technology was developed by IBM in the 1960's. CICS is a transaction processing system as well as a telecommunication system that can support many hundreds of terminals. For example, CICS was the enabling technology that supported the early automated teller machines (ATM).

CICS Elements
CICS applications are comprised of several elements, including a screen mapset, a program and a transaction. We'll discuss each of these broadly, and then we'll get into more detail in later chapters.

Screen Map
CICS applications consist of formatted screens that are an interface to CICS programs that provide some sort of processing, typically to display and update data in a data store. The screen component of CICS applications is usually developed using a CICS product called Basic Mapping Support (BMS).

To use BMS, you code instructions for how the screen is to be displayed, both the literal fields as well as user enterable fields. The instructions you provide to BMS will include the screen position and attributes of each field. Then you compile the instructions using a JCL (job control language) that produces what's called a mapset. A mapset contains one or more screen maps, and the map is what defines the screen that gets displayed.

As part of the assembly process, BMS will create a symbolic map which is a copybook that includes the layout of the screen including any enterable fields. The layout is used by the CICS program to receive data from the screen, as well as to send data back to the screen.

Here are some BMS terms you will need to know.

Map
A map is an individual screen format which you can create using BMS. The name of the mapset can be between 1 and 7 characters.

Mapset
A mapset is a set of maps which are defined and stored together to formulate a single load

module. The name of the mapset can be between 1 and 7 characters.

BMS Macros

BMS maps are written in Assembler language. Fortunately you don't need to be concerned with learning Assembler. Instead three macros are used to define a screen mapset. These macros are DFHMSD, DFHMDI, and DFHMDF. We'll give you the high level about these macros and then discuss them in more detail when we create some screens.

DFHMSD

DFHMSD defines a mapset. In the above case, we've named our mapset EMPMMNU.
You can have more than one map in a mapset, but most applications I've worked with use one map per mapset. When defining the mapset, you have these parameters:

TYPE

TYPE indicates the kind of map to be generated. There is both the physical map which defines how the screen looks and behaves. There is also a symbolic map which is a copybook that is used to pass data back and forth from the screen to the program. We'll get into that momentarily. Meanwhile, here are the possible values for TYPE:

```
MAP            define the physical map
DSECT          define the symbolic map
&&SYSPARM      define both the physical and symbolic map
```

Typically we will specify &&SYSPARM in our mapset definition because we want both the physical map and the symbolic map to be generated.

MODE

MODE indicates whether input and/or output operations are allowed. You can specify IN, OUT or INOUT. INOUT is usually specified.

CTRL

CTRL defines device control requests. You can specify the following:

```
FREEKB     Unlocks the keyboard.
FRSET      Resets the MDT to zero status.
ALARM      Sets an audible alarm at screen display time (if the
           device supports it).
PRINT      Causes the mapset to be printed out on a printer.
```

LANG

LANG indicates whether the program that uses the mapset is Assembler, COBOL, PLI or C language. The corresponding values are:

```
ASM
COBOL
PLI
C
```

TIOAPFX

TIOAPFX determines whether a filler space for BMS commands will be included in the symbolic map. The values are YES or NO.

DSATTS

DSATTS indicates which attributes to include in the symbolic description map. The valid values are:

```
COLOR
HILIGHT
OUTLINE
PS
SOSI
TRANSP
VALIDN
```

In our screen maps we will specify COLOR and HILITE.

MAPATTS

MAPATTS indicates which attributes to include in the physical map. The valid values are:

```
COLOR
HILIGHT
OUTLINE
PS
SOSI
TRANSP
VALIDN
```

In our screen maps we will specify COLOR and HILITE.

STORAGE

STORAGE indicates whether the symbolic maps in the mapset are to be defined with separate

15

storage areas, or whether they will use redefined storage space. Specifying STORAGE=AUTO means these maps will occupy separate storage areas.

Now let's actually look at a mapset definition. Here is the one we will use later for the EMP-MMNU mapset. We've indicated that the mapset name is EMPMMNU. We want it to generate both a physical and a symbolic map. The mapset is for a COBOL program to use. We want to be able to choose the COLOR of our fields and to set them to HILITE if desired.

```
----+----1----+----2----+----3----+----4----+----5----+----6----+----7----+----8
EMPMMNU  DFHMSD TYPE=&SYSPARM,MODE=INOUT,CTRL=(FREEKB,FRSET),          X
                LANG=COBOL,TIOAPFX=YES,                                X
                DSATTS=(COLOR,HILIGHT),                                X
                MAPATTS=(COLOR,HILIGHT),                               X
                STORAGE=AUTO
```

DFHMDI
The DFHMDI macro defines a screen map, and establishes that this is the beginning of the map. The map name can be up to 7 characters. Besides the map name you usually specify the number of lines and columns in the map. Typically this will be 24 lines and 80 columns. When defining the map, you have these parameters:

SIZE   The size of the map in lines and columns.

LINE   Specifies the starting line number of the map.

COLUMN   Specifies the starting column of the map.

JUSTIFY   Indicates the position of the map on the page. Valid values for this
                parameter are: LEFT, RIGHT, FIRST, LAST and BOTTOM

You can also specify CTRL and TIOAPFX at the map level, but we will not do that in our example. In fact, we'll simply specify the name and size for our EMPMNU map:

```
        EMPMNU   DFHMDI SIZE=(24,80)
```

DFHMDF
The DFHMDF macro defines fields and their attributes. Literal fields will not usually have tag names, but any field you want to reference in your program must have a tag name specified in positions 1-7. The name can be from 1 to 7 characters. I strongly suggest you create meaningful names, as these will be the variable names generated in the symbolic map (that your program will use).

16

When defining the map, you have these parameters:

POS
POS establishes the position of the new field in terms of the line and column number.

LENGTH
LENGTH specifies the length of the field.

INITIAL
INITIAL specifies the initial value for the field, if any.

JUSTIFY
JUSTIFY specifies whether the field is to be left or right justified.

ATTRB
ATTRB describes the attribute(s) for this field. Valid ATTRB values are as follows:

ASKIP
Autoskip, the cursor skips to the next field.

PROT
The field is protected. You cannot enter data here.

UNPROT
The field is unprotected. You can enter data here.

NUM
Only numeric data can be entered in this field.

BRT
Field is bright (highlighted).

NORM
Normal display (this is the default).

DRK
Dark display.

IC

Insert cursor (place the cursor on this field).

FSET
Field set, ensures data is sent from terminal to program even if nothing changes on the screen.

PICIN
Picture in – this describes the format of numeric data for an input field. For example PICIN = 9(8) describes an 8 digit numeric input.

PICOUT
Picture out – this describes the format of numeric data for an output field. For example PICOUT = 9(8) describes an 8 digit numeric output.

Sample Screen Map
Here's a sample of a screen we want to build.

```
EMPMMNU                    EMPLOYEE SUPPORT MENU                    EMNU

              ENTER THE NUMBER OF YOUR SELECTION,  THEN PRESS ENTER.


                     _    1. EMPLOYEE INQUIRY

                          2. EMPLOYEE ADD

                          3. EMPLOYEE CHANGE

                          4. EMPLOYEE DELETE




    F3 EXIT
```

Here is the BMS code to build the screen. I'm sure it looks somewhat cryptic. We'll explain each of the commands, tag names, etc.

```
----+----1----+----2----+----3----+----4----+----5----+----6----+----7----+----8
EMPMMNU   DFHMSD TYPE=&SYSPARM,MODE=INOUT,CTRL=(FREEKB,FRSET),          X
                 LANG=COBOL,TIOAPFX=YES,                               X
                 DSATTS=(COLOR,HILIGHT),                               X
                 MAPATTS=(COLOR,HILIGHT),                              X
                 STORAGE=AUTO
EMPMNU    DFHMDI SIZE=(24,80)
          DFHMDF POS=(01,1),LENGTH=07,COLOR=BLUE,                      X
                 INITIAL='EMPMMNU'
          DFHMDF POS=(01,31),LENGTH=21,COLOR=BLUE,                     X
                 INITIAL='EMPLOYEE SUPPORT MENU'
TRANID    DFHMDF POS=(01,76),LENGTH=04,INITIAL='EMNU',COLOR=BLUE
          DFHMDF POS=(03,15),LENGTH=35,COLOR=BLUE,                     X
                 INITIAL='ENTER THE NUMBER OF YOUR SELECTION,'
          DFHMDF POS=(03,52),LENGTH=17,COLOR=BLUE,                     X
                 INITIAL='THEN PRESS ENTER.'
ACTION    DFHMDF POS=(06,28),LENGTH=01,ATTRB=(IC,UNPROT),COLOR=GREEN,  X
                 HILIGHT=UNDERLINE
          DFHMDF POS=(06,30),LENGTH=01,ATTRB=ASKIP
          DFHMDF POS=(06,32),LENGTH=19,COLOR=BLUE,                     X
                 INITIAL='1. EMPLOYEE INQUIRY'
          DFHMDF POS=(08,32),LENGTH=15,COLOR=BLUE,                     X
                 INITIAL='2. EMPLOYEE ADD'
          DFHMDF POS=(10,32),LENGTH=18,COLOR=BLUE,                     X
                 INITIAL='3. EMPLOYEE CHANGE'
          DFHMDF POS=(12,32),LENGTH=18,COLOR=BLUE,                     X
                 INITIAL='4. EMPLOYEE DELETE'
MESSAGE   DFHMDF POS=(23,02),LENGTH=67,COLOR=YELLOW
          DFHMDF POS=(24,02),LENGTH=07,ATTRB=PROT,COLOR=BLUE,          X
                 INITIAL='F3 EXIT'
          DFHMSD TYPE=FINAL
          END
```

As you can see, we define the mapset using the DFHMSD macro and we name the mapset EMPMMNU. We specified that the mapset will be invoked by a program written in the COBOL language, and we provided a few other parameters.

Next we defined a map using the DFHMDI macro, and we named the map EMPMNU. We also specified the standard screen size of 24 lines of 80 bytes each.

Then we mapped how the screen will appear (both literal values and modifiable fields) with the DFHMDF macro. Note that tag names may be used for a field and if so the tag names begin in column 1.

As mentioned earlier, literal values such as the screen title do not require tag names, but input/output variables such as the action code do. For example we named the action code field ACTION. Then we specified by line number and column number exactly where the field is to be displayed on the screen (line 6, column 28).

We also specified each field's attributes. In this case we specified that the ACTION field is unprotected and the initial cursor position will be on this field. We further specified the color to be green, and that the field will be underlined to make it easy for the user to see.

Take a look at the rest of the fields that we defined, and reference their attributes back to the attribute list we provided earlier. This should give you a good idea what we are doing with the physical map.

Now let's take a look at the symbolic map that BMS generates for use in the application program..

Symbolic Map

The symbolic map is a copybook that includes 5 different fields for each input/output field. The variable names are the actual field name defined in the map, plus a suffix. For example the length of the field named ACTION is contained in a variable named ACTIONL. Here are the five suffixes and their usage:

L   Length variable that specifies the length of the entered data
F   Field variable indicates whether the field value has changed
I   Indicates an input variable - this is where the input data is stored
O   Indicates an output variable - this is where the output data is stored
A   Attribute variable that indicates the attributes for the field

In the screen map, we have a field named ACTION. We see below the corresponding variables are: ACTIONL, ACTIONF, ACTIONA, ACTIONI, ACTIONO

```
01   EMPMNUI.
     02  FILLER PIC X(12).
     02  TRANIDL    COMP PIC  S9(4).
     02  TRANIDF    PICTURE X.
     02  FILLER REDEFINES TRANIDF.
        03 TRANIDA     PICTURE X.
     02  FILLER   PICTURE X(2).
     02  TRANIDI PIC X(4).
     02  ACTIONL    COMP PIC  S9(4).
     02  ACTIONF    PICTURE X.
     02  FILLER REDEFINES ACTIONF.
        03 ACTIONA     PICTURE X.
     02  FILLER   PICTURE X(2).
```

```
02  ACTIONI  PIC X(1).
02  MESSAGEL    COMP PIC S9(4).
02  MESSAGEF    PICTURE X.
02  FILLER REDEFINES MESSAGEF.
   03 MESSAGEA    PICTURE X.
02  FILLER   PICTURE X(2).
02  MESSAGEI PIC X(67).
01  EMPMNUO REDEFINES EMPMNUI.
02  FILLER PIC X(12).
02  FILLER PICTURE X(3).
02  TRANIDC    PICTURE X.
02  TRANIDH    PICTURE X.
02  TRANIDO PIC X(4).
02  FILLER PICTURE X(3).
02  ACTIONC    PICTURE X.
02  ACTIONH    PICTURE X.
02  ACTIONO PIC X(1).
02  FILLER PICTURE X(3).
02  MESSAGEC    PICTURE X.
02  MESSAGEH    PICTURE X.
02  MESSAGEO PIC X(67).
```

Of course a screen map doesn't do anything without a program to do some processing. So let's move on to the CICS program.

Sample CICS Program

Our program will be named EMPPMNU, and the identification and environment divisions are simple.

```
IDENTIFICATION DIVISION.
PROGRAM-ID. EMPPGMNU.

***************************************************
*   MENU PROGRAM FOR EMPLOYEE APPLICATION       *
*                                               *
*   AUTHOR       : ROBERT WINGATE               *
*   DATE-WRITTEN : 2018-07-26                   *
***************************************************

ENVIRONMENT DIVISION.
```

Next, let's add the data division which will also be fairly simple.

```
DATA DIVISION.

WORKING-STORAGE SECTION.

01 WS-FLAGS.
   05 SW-VALID-SELECTION     PIC X(1) VALUE 'N'.
      88  VALID-SELECTION             VALUE 'Y'.
      88  NOT-VALID-SELECTION         VALUE 'N'.

01 WS-VARS.
```

```
   05 COMM-AREA                PIC X(20) VALUE SPACE.
   05 PROGRAM-NAME             PIC X(08) VALUE SPACES.
   05 INVALID-ACTION-MSG       PIC X(34)
      VALUE 'ENTER A VALID ACTION: 1, 2, 3 OR 4'.

    COPY EMPMMNU.
    COPY DFHAID.
    COPY DFHBMSCA.

 LINKAGE SECTION.

 01 DFHCOMMAREA          PIC X(20).
```

We've coded a flag to indicate if the user entered an invalid selection. We also define a communication area for data to be passed to other programs. We won't be passing data from this program but for consistency with the other programs we'll define a 20 byte are called COMM-AREA. Next, we've defined an 8 byte variable which will contain the name of the program we will transfer to if the user requests it. We'll also define a message literal that will be used when the user enters an invalid selection.

We're also including three copybooks as follows:

1. **EMPMMNU** is the symbolic map generated when we compile the mapset.

2. **DFHAID** contains the standard attention identifier list which is a set of literal names for the various key presses that are captured. For example DFHPF3 indicates that the PF3 key was pressed. DFHENTER means that the ENTER key was pressed. The complete list is on the IBM product support web site.

https://www.ibm.com/support/knowledgecenter/en/SSAL2T_8.1.0/com.ibm.cics.tx.doc/reference/r_attn_idntfr_consts_lst.html

3. **DFHBMSCA** contains constants for setting various values such as attribute characters. For example DFHBMFSE is a constant that means the value of a field has changed. For example, in our symbolic map the field ACTIONF is the attribute byte for the ACTIONI field. If the value of ACTIONF is equal to DFHBMFSE, it means the field value has changed. The complete list is on the IBM product support web site.

https://www.ibm.com/support/knowledgecenter/en/SSGMCP_5.3.0/com.ibm.cics.ts.applicationprogramming.doc/topics/dfhp4_bmsconstants.html

Finally, in the linkage section we must define a variable named DFHCOMMAREA. This is where another program can pass data to this program. We won't be doing that in the menu program, but again we want consistency to the other programs so we will define DFHCOM-MAREA as character 20 bytes.

For the procedure division, let's divide our discussion into three parts:

1. The sending and receiving of data using the symbolic map.
2. The checking for keys pressed by the user.
3. The actual processing of the user's request.

In any transactional system there must be commands to retrieve the user's screen input, and to send output back to the screen. In CICS the commands are SEND and RECEIVE used with various parameters. To erase the screen and send the menu screen map, you would use the following command:

```
EXEC CICS SEND
    MAP    ('EMPMNU')
    MAPSET ('EMPMMNU')
    FROM   (EMPMNUO)
    ERASE
END-EXEC.
```

This directs CICS to send the EMPMNU map located in the EMPMMNU mapset to the user's terminal, and to populate the variables with data from the symbolic map EMPMNUO. The latter is the output map from the copybook EMPMMNU that was generated when we compiled the screen. We'll go through the steps in more detail in the coming chapters. Also notice that we specified the ERASE parameter to remove any residual screen data from a previous screen.

Once the user has entered a selection and pressed the Enter key, we will receive the data into the program as follows:

```
EXEC CICS RECEIVE
    MAP    ('EMPMNU')
    MAPSET ('EMPMMNU')
    INTO   (EMPMNUI)
END-EXEC.
```

Here we command CICS to receive screen data using the specified map and mapset, and we indicate the symbolic input map as a container for the data.

Finally, once the program has processed the user's input, the program will send the map back to the screen where it will be displayed. Typically in this case, you only send the modified data back (not the entire screen) by specifying the DATAONLY parameter.

```
EXEC CICS SEND
    MAP     ('EMPMNU')
    MAPSET  ('EMPMMNU')
    FROM    (EMPMNUO)
    DATAONLY
END-EXEC.
```

So let's program three procedures that use the SEND and RECEIVE commands as follows:

```
SEND-MAP.
    EXEC CICS SEND
        MAP     ('EMPMNU')
        MAPSET  ('EMPMMNU')
        FROM    (EMPMNUO)
        ERASE
    END-EXEC.

SEND-MAP-DATAONLY.
    EXEC CICS SEND
        MAP     ('EMPMNU')
        MAPSET  ('EMPMMNU')
        FROM    (EMPMNUO)
        DATAONLY
    END-EXEC.


RECEIVE-MAP.
    EXEC CICS RECEIVE
        MAP     ('EMPMNU')
        MAPSET  ('EMPMMNU')
        INTO    (EMPMNUI)
    END-EXEC.
```

Ok, next we want to code the program mainline to intercept and handle whatever conditions or keyed input occur. Here is our code below, and we'll explain each element.

There is a variable named EIBCALEN that indicates how many times this session has been through the program. If the value is zero, that means this is the first time through. If it is the first time through, we simply want to display the screen for the user. So in the code below, we initialize the symbolic output map, and then we call the send map procedure.

Next, we check to see if the user pressed the clear key, the PA keys, or the PF3 key and handle those as indicated. Next, if the user pressed the ENTER key, we call the procedure to process the user's selection. Finally, if the user pressed any other key, we set up the "invalid key pressed" error message and then send the map.

```
IF EIBCALEN > ZERO
  MOVE DFHCOMMAREA TO COMM-AREA
END-IF.

EVALUATE TRUE

  WHEN EIBCALEN = ZERO
    MOVE LOW-VALUES    TO   EMPMNUO
    PERFORM SEND-MAP

  WHEN EIBAID = DFHCLEAR
    MOVE LOW-VALUES    TO   EMPMNUO
    PERFORM SEND-MAP

  WHEN EIBAID = DFHPA1 OR DFHPA2 OR DFHPA3
    CONTINUE

  WHEN EIBAID = DFHPF3
    MOVE LOW-VALUES TO  EMPMNUO
    MOVE "BYE, PRESS CLEAR KEY TO ENTER A TRANSACTION ID"
         TO MESSAGEO
    PERFORM SEND-MAP-DATAONLY

    EXEC CICS
      RETURN
    END-EXEC

  WHEN EIBAID = DFHENTER
    PERFORM MAIN-PROCESS-PARA

  WHEN OTHER
    MOVE LOW-VALUES TO EMPMNUO
    MOVE "INVALID KEY PRESSED" TO MESSAGEO
    PERFORM SEND-MAP-DATAONLY

END-EVALUATE.

EXEC CICS
   RETURN TRANSID('EMNU')
   COMMAREA (COMM-AREA)
END-EXEC.
```

Ok, at this point we've handled everything except processing the user's request. To do that, we will need two procedures. One is the MAIN-PROCESS-PARA which will evaluate the user's request and call the appropriate program. The other is the CICS routine to transfer control to the requested program. Here's the main process routine. Basically what we're doing is to:

1. Receive the screen map.
2. Check for a valid action number.

3. If valid, load the program name and call the branch-to procedure
4. If invalid, load the error message
5. . Send the map with data only

Here's the main processing procedure:

```
MAIN-PROCESS-PARA.

    PERFORM RECEIVE-MAP.

    IF ACTIONI NOT = '1' AND '2' AND '3' AND '4'
       MOVE DFHREVRS TO ACTIONH
       MOVE INVALID-ACTION-MSG TO MESSAGEO
       SET NOT-VALID-SELECTION TO TRUE
    ELSE
       SET VALID-SELECTION TO TRUE
    END-IF.

    IF VALID-SELECTION
       EVALUATE ACTIONI
          WHEN '1'
             MOVE 'EMPPGINQ' TO PROGRAM-NAME
          WHEN '2'
             MOVE 'EMPPGADD' TO PROGRAM-NAME
          WHEN '3'
             MOVE 'EMPPGCHG' TO PROGRAM-NAME
          WHEN '4'
             MOVE 'EMPPGDEL' TO PROGRAM-NAME
       END-EVALUATE

       PERFORM BRANCH-TO-PROGRAM

    END-IF.

    PERFORM SEND-MAP-DATAONLY.
```

In our BRANCH-TO-PROGRAM paragraph we use the CICS XCTL command to transfer control to the specified program. We just need to provide the program name which we do using the variable PROGRAM-NAME.

```
BRANCH-TO-PROGRAM.

    EXEC CICS
       XCTL PROGRAM(PROGRAM-NAME)
    END-EXEC

    MOVE 'PROGRAM NOT AVAILABLE' TO MESSAGEO.
```

Ok, that's it. Here is the complete program code. If some of this doesn't make 100% sense, don't worry. We'll explain more later. This program example is just for illustration now so you get a taste of CICS programming.

```
    IDENTIFICATION DIVISION.
    PROGRAM-ID. EMPPGMNU.

***************************************************
*  MENU PROGRAM FOR EMPLOYEE APPLICATION          *
*                                                 *
*  AUTHOR       : ROBERT WINGATE                   *
*  DATE-WRITTEN : 2018-07-26                       *
***************************************************

    ENVIRONMENT DIVISION.

    DATA DIVISION.

    WORKING-STORAGE SECTION.

    01 WS-FLAGS.
       05 SW-VALID-SELECTION      PIC X(1) VALUE 'N'.
          88  VALID-SELECTION              VALUE 'Y'.
          88  NOT-VALID-SELECTION          VALUE 'N'.

    01 WS-VARS.
       05 COMM-AREA              PIC X(20) VALUE SPACE.
       05 PROGRAM-NAME           PIC X(08) VALUE SPACES.
       05 INVALID-ACTION-MSG     PIC X(34)
          VALUE 'ENTER A VALID ACTION: 1, 2, 3 OR 4'.

       COPY EMPMMNU.
       COPY DFHAID.
       COPY DFHBMSCA.

    LINKAGE SECTION.

    01 DFHCOMMAREA         PIC X(20).

    PROCEDURE DIVISION.

       IF EIBCALEN > ZERO
         MOVE DFHCOMMAREA  TO COMM-AREA
       END-IF.

       EVALUATE TRUE

         WHEN EIBCALEN = ZERO
           MOVE LOW-VALUES   TO EMPMNUO
           PERFORM SEND-MAP

         WHEN EIBAID = DFHCLEAR
           MOVE LOW-VALUES    TO EMPMNUO
           PERFORM SEND-MAP

         WHEN EIBAID = DFHPA1 OR DFHPA2 OR DFHPA3
           CONTINUE

         WHEN EIBAID = DFHPF3
           MOVE LOW-VALUES TO  EMPMNUO
           MOVE "BYE, PRESS CLEAR KEY TO ENTER A TRANSACTION ID"
```

27

```
                 TO MESSAGEO
           PERFORM SEND-MAP-DATAONLY

           EXEC CICS
             RETURN
           END-EXEC

        WHEN EIBAID = DFHENTER
           PERFORM MAIN-PROCESS-PARA

        WHEN OTHER
           MOVE LOW-VALUES TO EMPMNUO
           MOVE "INVALID KEY PRESSED" TO MESSAGEO
           PERFORM SEND-MAP-DATAONLY

     END-EVALUATE.

     EXEC CICS
        RETURN TRANSID('EMNU')
        COMMAREA (COMM-AREA)
     END-EXEC.

MAIN-PROCESS-PARA.

     PERFORM RECEIVE-MAP.

     IF ACTIONI NOT = '1' AND '2' AND '3' AND '4'
        MOVE DFHREVRS TO ACTIONH
        MOVE INVALID-ACTION-MSG TO MESSAGEO
        SET NOT-VALID-SELECTION TO TRUE
     ELSE
        SET VALID-SELECTION TO TRUE
     END-IF.

     IF VALID-SELECTION
        EVALUATE ACTIONI
           WHEN '1'
              MOVE 'EMPPGINQ' TO PROGRAM-NAME
           WHEN '2'
              MOVE 'EMPPGADD' TO PROGRAM-NAME
           WHEN '3'
              MOVE 'EMPPGCHG' TO PROGRAM-NAME
           WHEN '4'
              MOVE 'EMPPGDEL' TO PROGRAM-NAME
        END-EVALUATE

        PERFORM BRANCH-TO-PROGRAM

     END-IF.

     PERFORM SEND-MAP-DATAONLY.

BRANCH-TO-PROGRAM.

     EXEC CICS
        XCTL PROGRAM(PROGRAM-NAME)
     END-EXEC
```

```
          MOVE 'PROGRAM NOT AVAILABLE' TO MESSAGEO.

      SEND-MAP.
          EXEC CICS SEND
              MAP     ('EMPMNU')
              MAPSET  ('EMPMMNU')
              FROM    (EMPMNUO)
              ERASE
          END-EXEC.

      SEND-MAP-DATAONLY.
          EXEC CICS SEND
              MAP     ('EMPMNU')
              MAPSET  ('EMPMMNU')
              FROM    (EMPMNUO)
              DATAONLY
          END-EXEC.


      RECEIVE-MAP.
          EXEC CICS RECEIVE
              MAP     ('EMPMNU')
              MAPSET  ('EMPMMNU')
              INTO    (EMPMNUI)
          END-EXEC.
```

Sample Transaction

You invoke a CICS program through a transaction. Transaction identifiers can be between 1 and 4 characters long. All shops I've ever worked in use 4 characters by convention. So we could define transaction EMNU for our menu program, and that is how the transaction is invoked (by typing EMNU on a blank CICS screen and pressing ENTER)

In order to make these components active, you must define your mapset, program and transaction to CICS. This is typically done using the CEDA utility. We will do that later in another chapter. For now, we've seen examples of a map, mapset and program, and we've discussed at a high level that these are linked using a transaction.

In the next chapter, we will provide specifications for a simple employee support application to be used by a fictitious HR department. This will give us something structured to work with as we explain more complex CICS concepts and start to do actual programming.

Chapter Two: Employee Support Application Design
Purpose
This design is for a Human Resource application called Employee Support. It consists of a data store containing employee information such as employee number, name, years of service and social security number. There will be several CICS screen that allow the user to:

1. Select from a menu of options (Main Menu)
2. Display employee detail for a specified employee
3. Add detail for a new employee
4. Change detail for an existing employee
5. Delete detail for an employee

General Design
The following provides a high level design of the application in terms of data store, screen, program and transaction components.

Data Design
The following table specification is provided:

EMPLOYEE (key is EMP_ID).

Field Name	Type
EMP_ID	INTEGER
EMP_LAST_NAME	VARCHAR(30)
EMP_FIRST_NAME	VARCHAR(20)
EMP_SERVICE_YEARS	INTEGER
EMP_PROMOTION_DATE	DATE
EMP_SSN	VARCHAR(09)

The fields are self explanatory except for EMP_SSN which is the employee's social security number.

Application Elements

Screen Components
Here we specify the screen mapset names and the transactions that will be associated with them.

Screen Name	Trans ID	Mapset Name
Employee Support Menu	EMNU	EMPMMNU
Employee Inquiry	EMIN	EMPMINQ
Employee Add	EMAD	EMPMADD
Employee Change	EMCH	EMPMCHG
Employee Delete	EMDE	EMPMDEL

Hierarchy Chart
This is a simple functional hierarchy chart. It indicates that the menu program will call one of four different programs: the Employee Inquiry, Employee Add, Employee Change and Employee Delete.

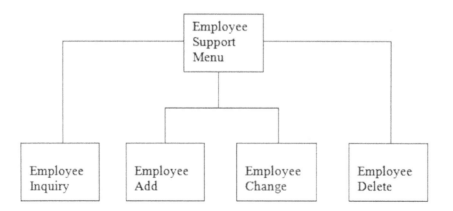

Program Elements (COBOL)

The following table summarizes the program components required for our application and the transactions that will invoke the programs.

Screen Name	Trans ID	Program Name
Employee Support Menu	EMNU	EMPPGMNU
Employee Inquiry	EMIN	EMPPGINQ
Employee Add	EMAD	EMPPGADD
Employee Change	EMCH	EMPPGCHG
Employee Delete	EMDE	EMPPGDEL

Detail Design
Detailed Data Design
In this section we will design and create the appropriate DB2 table for our application. You'll need to either have the DBA assign you a database to work with or create one yourself. If you have authority, I suggest you create a Human Resource database named DBHR. Also you can create a tablespace called TSHR, and a schema called HRSCHEMA.[1]

The following is just sample DDL. You must know your system in order to supply the correct values.

```
CREATE DATABASE DBHR
STOGROUP SGHR
BUFFERPOOL BPHR
INDEXBP IBPHR
CCSID UNICODE;

CREATE TABLESPACE TSHR
IN DBHR
USING STOGROUP SGHR
PRIQTY 50
SECQTY 20
LOCKSIZE PAGE
BUFFERPOOL BPHR2;

CREATE SCHEMA HRSCHEMA
AUTHORIZATION USER01;    ←    This should be your DB2 id, whatever it is.
```

Now you can create the EMPLOYEE table. Here is the DDL:

```
CREATE TABLE HRSCHEMA.EMPLOYEE(
EMP_ID INT NOT NULL,
EMP_LAST_NAME VARCHAR(30) NOT NULL,
EMP_FIRST_NAME VARCHAR(20) NOT NULL,
EMP_SERVICE_YEARS INT NOT NULL WITH DEFAULT 0,
EMP_PROMOTION_DATE DATE,
EMP_SSN  CHAR(09),
PRIMARY KEY(EMP_ID));
```

We also need to create a unique index to support the primary key:

```
CREATE UNIQUE INDEX NDX_EMPLOYEE
ON EMPLOYEE (EMP_ID);
```

---

1 If you are not familiar with DB2 databases, tablespaces and schema you can check out Quick Start Training for IBM z/OS Application Developers Volume 1. It includes a chapter on basic DB2. You can also choose DB2 for z/OS Basic Training for Application Developers. These are both listed in the back of this book.

Now let's insert some data.

```
INSERT INTO HRSCHEMA.EMPLOYEE
VALUES (3217,
'JOHNSON',
'EDWARD',
4,
'01/01/2017',
'397342007');

INSERT INTO HRSCHEMA.EMPLOYEE
VALUES (7459,
'STEWART',
'BETTY',
7,
'07/31/2016',
' 019572830');


INSERT INTO HRSCHEMA.EMPLOYEE
VALUES (9134,
'FRANKLIN',
'BRIANNA',
DEFAULT,
NULL,
' 937293598');


INSERT INTO HRSCHEMA.EMPLOYEE
VALUES (4720,
'SCHULTZ',
'TIM',
9,
'01/01/2017',
' 650450254');


INSERT INTO HRSCHEMA.EMPLOYEE
VALUES (6288,
'WILLARD',
'JOE',
6,
'01/01/2016',
' 209883920');
```

Ok that's enough data to start with. We'll use it when we get to the actual programming.

CICS Screen Designs

Employee Support Menu
This screen allows the user to select one of four options:

1. Employee Inquiry
2. Employee Add
3. Employee Change
4. Employee Delete

The following are the transaction, mapset and map names we will use for this program.

```
Transaction Name: EMNU
Screen Mapset:    EMPMMNU
Program Name:     EMPPGMNU
Data Stores:      None
```

The display should be as follows. We'll see later that there is a message field just above the F3 EXIT display. For now this field is empty, so it does not show any message. It will be used when the program needs to display an error message, or in some cases to prompt the user for an additional action.

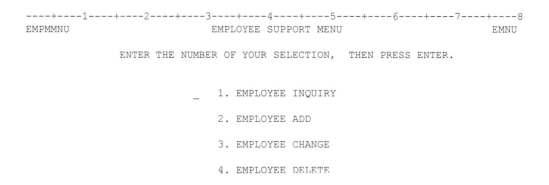

```
----+----1----+----2----+----3----+----4----+----5----+----6----+----7----+----8
EMPMMNU                     EMPLOYEE SUPPORT MENU                      EMNU

           ENTER THE NUMBER OF YOUR SELECTION,   THEN PRESS ENTER.


                        _   1. EMPLOYEE INQUIRY

                            2. EMPLOYEE ADD

                            3. EMPLOYEE CHANGE

                            4. EMPLOYEE DELETE
```

```
F3 EXIT
```

Note: The numbers on the top line of the display are just for reference – they are not part of the actual screen image.

The user types the number of the desired selection in the input field which is the underscore to the left of the first choice (item 1). So the user enters 1, 2 3, or 4 and then presses ENTER. The result is that the selected screen will display.

Employee Inquiry

This screen allows a user to enter an employee id. After doing so and pressing ENTER, the detailed information for the employee will be displayed. If the requested employee id is not in the EMPLOYEE table, then an error message will be returned. The user will also be able to switch to a different option such as the add, change or delete screen.

```
Transaction Name: EMIN
Screen Mapset:    EMPMINQ
Program Name:     EMPPGINQ
Data Stores:      EMPLOYEE
```

The display should be as follows. The user enters a valid employee id and presses ENTER. If the employee is found in the table, the detailed employee information is displayed on the screen. If the employee is not found, an error message is displayed.

```
----+----1----+----2----+----3----+----4----+----5----+----6----+----7----+----8
EMPMINQ                     EMPLOYEE INQUIRY                            EMIN

        EMPLOYEE ->     ____        ENTER EMPLOYEE ID, THEN PRESS ENTER

        EMPLOYEE ID

        EMP LAST NAME

        EMP FIRST NAME

        EMP SOCIAL SEC

        EMP YEARS SRVC

        EMP LAST PROM




   F2 INQ   F3 EXIT   F4 ADD   F5 CHG   F6 DEL
```

For example, if the user enters employee 3217, and presses the ENTER key, the following should result:

```
----+----1----+----2----+----3----+----4----+----5----+----6----+----7----+----8

EMPMINQ                    EMPLOYEE INQUIRY                          EMIN


        EMPLOYEE ->     3217        ENTER EMPLOYEE ID, THEN PRESS ENTER

        EMPLOYEE ID    3217

        EMP LAST NAME  JOHNSON

        EMP FIRST NAME EDWARD

        EMP SOCIAL SEC 397342007

        EMP YEARS SRVC 04

        EMP LAST PROM  2017-01-01




  F2 INQ   F3 EXIT   F4 ADD   F5 CHG   F6 DEL
```

If the user enters an invalid employee id, an error message should result.

```
----+----1----+----2----+----3----+----4----+----5----+----6----+----7----+----8
EMPMINQ                    EMPLOYEE INQUIRY                          EMIN

        EMPLOYEE ->     3188        ENTER EMPLOYEE ID, THEN PRESS ENTER

        EMPLOYEE ID

        EMP LAST NAME

        EMP FIRST NAME

        EMP SOCIAL SEC

        EMP YEARS SRVC

        EMP LAST PROM
```

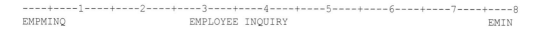

```
EMPLOYEE ID 3188 NOT FOUND
F2 INQ   F3 EXIT   F4 ADD   F5 CHG   F6 DEL
```

Employee Add

This screen allows a user to enter a new employee. The detailed information for the employee will be added to the EMPLOYEE table. The user will also be able to switch to a different option such as the inquiry, change or delete screen.

```
Transaction Name: EMAD
Screen Mapset:    EMPMADD
Program Name:     EMPPGADD
Data Stores:      EMPLOYEE
```

The initial display should be as follows. The user enters an employee id, last and first names, years of service, last promotion date and social security number. If the record is successfully added, a message will indicate this.

```
----+----1----+----2----+----3----+----4----+----5----+----6----+----7----+----8
EMPMADD                   EMPLOYEE ADD                            EMAD


                  ENTER EMPLOYEE INFO, THEN PRESS PF4

         EMPLOYEE ID        _____

         EMP LAST NAME      _____

         EMP FIRST NAME     _____

         EMP SOCIAL SEC     _____

         EMP YEARS SRVC     __

         EMP LAST PROM      _____




 ENTER DATA FOR NEW EMPLOYEE, THEN PRESS PF4 TO ADD
 F2 INQ   F3 EXIT   F4 ADD   F5 CHG   F6 DEL
```

Suppose we add employee 8888 with name Joan Sanders with social security number 432993928, 5 years of service and last promotion date 2018-07-01. Here is the screen filled in:

```
EMPMADD                          EMPLOYEE ADD                          EMAD


        EMPLOYEE ->    ENTER EMPLOYEE INFO, THEN PRESS PF4PRESS ENTER

        EMPLOYEE ID    8888

        EMP LAST NAME  Sanders

        EMP FIRST NAME Joan

        EMP SOCIAL SEC 432993928

        EMP YEARS SRVC 05

        EMP LAST PROM  2018-07-01




     ENTER DATA FOR NEW EMPLOYEE, THEN PRESS PF4 TO ADD
     F2 INQ   F3 EXIT   F4 ADD   F5 CHG   F6 DEL
```

Now press PF4. If successful you should see this screen with message indicating successfully added.

```
EMPMADD                          EMPLOYEE ADD                          EMAD


        EMPLOYEE ->    ENTER EMPLOYEE INFO, THEN PRESS PF4PRESS ENTER

        EMPLOYEE ID    8888

        EMP LAST NAME  SANDERS

        EMP FIRST NAME JOAN

        EMP SOCIAL SEC 432993928

        EMP YEARS SRVC 05

        EMP LAST PROM  2018-07-01




     EMPLOYEE ADDED SUCCESSFULLY
     F2 INQ   F3 EXIT   F4 ADD   F5 CHG   F6 DEL
```

Edits and Validations

We haven't yet specified the edits and validations for our business design. The fields must be checked for conformance with the following rules:

1. Employee number is required and must be numeric and greater than zero.
2. Last name is required.
3. First name is required.
4. Social Security Number is required and must be numeric and greater than zero.
5. Year of Service is required and must be numeric and greater than zero.
6. Last promotion date must be a valid date value in format YYYY-MM-DD.

Employee Change

This screen allows a user to enter an employee id, press ENTER and the detailed information for the employee will be returned. The user is prompted to make changes and then press PF5 to apply the changes to the table.

```
Transaction Name: EMCH
Screen Mapset:    EMPMCHG
Program Name:     EMPPGCHG
Data Stores:      EMPLOYEE
```

```
----+----1----+----2----+----3----+----4----+----5----+----6----+----7----+----8
EMPMCHG                   EMPLOYEE CHANGE                             EMCH


        EMPLOYEE ->      ____        ENTER EMPLOYEE ID, THEN PRESS ENTER

        EMPLOYEE ID

        EMP LAST NAME

        EMP FIRST NAME

        EMP SOCIAL SEC

        EMP YEARS SRVC

        EMP LAST PROM




  F2 INQ   F3 EXIT   F4 ADD   F5 CHG   F6 DEL
```

For example, suppose you want to change the years of service for employee 3217 from 4 year to 5 years. Here's the screen with the data pulled up:

```
EMPMCHG                    EMPLOYEE CHANGE                         EMCH


        EMPLOYEE ->    3217      ENTER EMPLOYEE ID, THEN PRESS ENTER

        EMPLOYEE ID    3217

        EMP LAST NAME  JOHNSON

        EMP FIRST NAME EDWARD

        EMP SOCIAL SEC 397342007

        EMP YEARS SRVC 04

        EMP LAST PROM  2017-01-01




    MAKE CHANGES AND THEN PRESS PF5
    F2 INQ   F3 EXIT   F4 ADD   F5 CHG   F6 DEL
```

If you make the change and press PF5, and the change is successful, you will receive a confirmation message that the change was successful.

```
EMPMCHG                    EMPLOYEE CHANGE                         EMCH


        EMPLOYEE ->    3217      ENTER EMPLOYEE ID, THEN PRESS ENTER

        EMPLOYEE ID    3217

        EMP LAST NAME  JOHNSON

        EMP FIRST NAME EDWARD

        EMP SOCIAL SEC 397342007

        EMP YEARS SRVC 05

        EMP LAST PROM  2017-01-01




    EMPLOYEE MODIFIED SUCCESSFULLY
    F2 INQ   F3 EXIT   F4 ADD   F5 CHG   F6 DEL
```

If the user presses ENTER instead of the PF5 key, they will continue to receive the confirmation message to change the record. Of course the user can also press another PF key to take another action such as switching to another screen or exiting.

Employee Delete

This screen allows a user to enter an employee id, press ENTER and the detailed information for the employee will be returned. The user will then be prompted to delete the employee from the table by pressing PF5.

```
Transaction Name: EMDE
Screen Mapset:    EMPMDEL
Program Name:     EMPPGDEL
Data Stores:      EMPLOYEE
```

The initial display should be as follows.

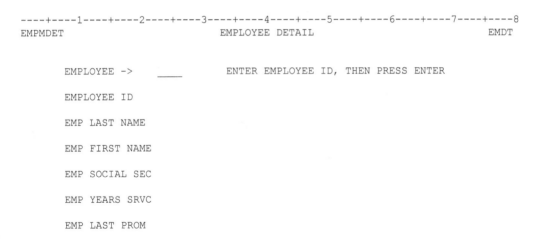

```
----+----1----+----2----+----3----+----4----+----5----+----6----+----7----+----8
EMPMDET                         EMPLOYEE DETAIL                         EMDT

         EMPLOYEE ->      ____        ENTER EMPLOYEE ID, THEN PRESS ENTER

         EMPLOYEE ID

         EMP LAST NAME

         EMP FIRST NAME

         EMP SOCIAL SEC

         EMP YEARS SRVC

         EMP LAST PROM




   F2 INQ    F3 EXIT    F4 ADD    F5 CHG    F6 DEL
```

Once the employee information is returned to the screen the user will be prompted by a message to press PF6 to delete the record. If the user presses ENTER instead of the PF6 key, they will continue to receive the confirmation message to delete the record. Of course the user can also press another PF key to take another action such as switching to another screen or exiting.

```
EMPMDEL                    EMPLOYEE DELETE                          EMDE


     EMPLOYEE ->    8888        ENTER EMPLOYEE ID, THEN PRESS ENTER

     EMPLOYEE ID    8888

     EMP LAST NAME  WINGATE

     EMP FIRST NAME ROBERT

     EMP SOCIAL SEC 454084595

     EMP YEARS SRVC 05

     EMP LAST PROM  2016-01-01




PRESS PF6 TO DELETE EMPLOYEE
F2 INQ   F3 EXIT   F4 ADD   F5 CHG   F6 DEL
```

If they do press PF6 they will receive a confirmation message that the record was deleted.

```
EMPMDEL                    EMPLOYEE DELETE                          EMDE


     EMPLOYEE ->    8888        ENTER EMPLOYEE ID, THEN PRESS ENTER

     EMPLOYEE ID    8888

     EMP LAST NAME  WINGATE

     EMP FIRST NAME ROBERT

     EMP SOCIAL SEC 454084595

     EMP YEARS SRVC 05

     EMP LAST PROM  2016-01-01




EMPLOYEE DELETED SUCCESSFULLY
F2 INQ   F3 EXIT   F4 ADD   F5 CHG   F6 DEL
```

Ok this completes the business design of our CICS screens. In the next chapter we will write and compile the code to create the screens.

Chapter Three: Employee Support Construction

BMS Coding Specifications and Assembly

BMS coding can be rather cryptic until you've done a few screens. We'll start out with the menu screen and explain how we create it. Then we'll jump into CICS to display it to make sure we have it correct.

Coding the EMPMMNU Screen

The BMS code for our menu screen is as follows. Here are a couple of general things to keep in mind. There are various types of entries, but almost all of them describe what goes on the screen and the attributes associated with those fields. Take a good look at the code and then we'll go through the basics of compiling, defining and installing the screen mapset.

```
----+----1----+----2----+----3----+----4----+----5----+----6----+----7----+----8
EMPMMNU   DFHMSD TYPE=&SYSPARM,MODE=INOUT,CTRL=(FREEKB,FRSET),         X
                LANG=COBOL,TIOAPFX=YES,                               X
                DSATTS=(COLOR,HILIGHT),                               X
                MAPATTS=(COLOR,HILIGHT),                              X
                STORAGE=AUTO
EMPMNU    DFHMDI SIZE=(24,80)
          DFHMDF POS=(01,1),LENGTH=07,COLOR=BLUE,                     X
                INITIAL='EMPMMNU'
          DFHMDF POS=(01,31),LENGTH=21,COLOR=BLUE,                    X
                INITIAL='EMPLOYEE SUPPORT MENU'
TRANID    DFHMDF POS=(01,76),LENGTH=04,INITIAL='EMNU',COLOR=BLUE
          DFHMDF POS=(03,15),LENGTH=35,COLOR=BLUE,                    X
                INITIAL='ENTER THE NUMBER OF YOUR SELECTION,'
          DFHMDF POS=(03,52),LENGTH=17,COLOR=BLUE,                    X
                INITIAL='THEN PRESS ENTER.'
ACTION    DFHMDF POS=(06,28),LENGTH=01,ATTRB=(IC,UNPROT),COLOR=GREEN, X
                HILIGHT=UNDERLINE
          DFHMDF POS=(06,30),LENGTH=01,ATTRB=ASKIP
          DFHMDF POS=(06,32),LENGTH=19,COLOR=BLUE,                    X
                INITIAL='1. EMPLOYEE INQUIRY'
          DFHMDF POS=(08,32),LENGTH=15,COLOR=BLUE,                    X
                INITIAL='2. EMPLOYEE ADD'
          DFHMDF POS=(10,32),LENGTH=18,COLOR=BLUE,                    X
                INITIAL='3. EMPLOYEE CHANGE'
          DFHMDF POS=(12,32),LENGTH=18,COLOR=BLUE,                    X
                INITIAL='4. EMPLOYEE DELETE'
MESSAGE   DFHMDF POS=(23,02),LENGTH=67,COLOR=YELLOW
          DFHMDF POS=(24,02),LENGTH=07,ATTRB=PROT,COLOR=BLUE,         X
                INITIAL='F3 EXIT'
          DFHMSD TYPE=FINAL
          END
```

Columns 1 – 8 are tag names that you can refer to in your program, including the mapset name, the map name, and any enterable fields you include to allow input and output. We'll talk about those momentarily.

Column 10 includes a keyword that begins an entry. These are the keywords that are normally required to define a map.

`DFHMSD` – this is the mapset name.

`DFHMDI` – this is the map name. You can include more than one map in a mapset. I have never worked in a shop that did this, but you can do it.

`DFHMDF` – this is where you actually define what goes on the screen.

```
DFHMSD TYPE=FINAL
END
```

The latter terminates the mapset.

Column 17 specifies detail about the entry just given. If the specification requires more than one line, a continuation character X must be placed in column 72. You can then continue the specification beginning in column 16 on the next line. Here's an example from the map:

```
ACTION    DFHMDF POS=(06,28),LENGTH=01,ATTRB=(IC,UNPROT),COLOR=GREEN,    X
               HILIGHT=UNDERLINE
```

The above created the input/output field called `ACTION` on line 6, column 28.

Compiling the EMPMMNU Screen

To generate the screen map, you must assemble it using the BMS assembler program. Henceforth in this book, we'll refer to this as compiling the mapset or screen. While technically we are "assembling it", compile is a more commonly used term for making a program executable, so we'll use it.

You'll need to locate the appropriate compile JCL used in your shop. I will show you the one I use, but it will undoubtedly be different that yours. Check with your supervisor or technical leader for your shop's JCL.

```
//USER01D JOB 'ASSEMBLE MAP',CLASS=A,
//              MSGLEVEL=(1,1),NOTIFY=&SYSUID
//*
//* TO ASSEMBLE CICS BMS MAP
//*
//CICSMAP  EXEC DFHMAPS,
//           COPYLIB=USER01.COPYLIB,       <= SYMBOLIC MAP COPY LIBRARY
//           SRCLIB=USER01.CICS.MAPLIB,    <= SOURCE LIBRARY
//           MEMBER=EMPMMNU                <= MAP MEMBER NAME
//*
```

Always check the output for errors. The most common BMS error I've seen is overlapping fields. A good compile will have these words near the end of the output:

```
No Statements Flagged in this Assembly
```

Using CEDA to Define and Install Components

Before you can use a CICS resource you must define and install it in CICS. CEDA is a CICS transaction that allows a developer or administrator to add and maintain resource definitions. To bring up CEDA, logon to CICS and enter transaction CEDA. You will see the following:

```
ENTER ONE OF THE FOLLOWING

ADd
ALter
APpend
CHeck
COpy
DEFine
DELete
DIsplay
Expand
Install
Lock
Move
REMove
REName
UNlock
USerdefine
View
                                          SYSID=CICS APPLID=CICSTS42


PF 1 HELP      3 END          6 CRSR         9 MSG         12 CNCL
```

Defining a Mapset

Type DEF or DEFINE and press enter. You will see this screen.

```
DEF
 ENTER ONE OF THE FOLLOWING

Atomservice   MQconn        Webservice
Bundle        PARTItionset
CONnection    PARTNer
CORbaserver   PIpeline
DB2Conn       PROCesstype
DB2Entry      PROFile
DB2Tran       PROGram
DJar          Requestmodel
DOctemplate   Sessions
Enqmodel      TCpipservice
File          TDqueue
Ipconn        TErminal
JOurnalmodel  TRANClass
JVmserver     TRANSaction
LIbrary       TSmodel
LSrpool       TYpeterm
MApset        Urimap


                                         SYSID=CICS APPLID=CICSTS42


PF 1 HELP      3 END         6 CRSR        9 MSG        12 CNCL
```

Now type MAPSET and you will see this screen:

```
DEF mapset
 OVERTYPE TO MODIFY                              CICS RELEASE = 0670
  CEDA  DEFine MApset(         )
   MApset        ==>
   Group         ==>
   DEScription   ==>
   REsident      ==> No             No | Yes
   USAge         ==> Normal         Normal | Transient
   USElpacopy    ==> No             No | Yes
   Status        ==> Enabled        Enabled | Disabled
   RSl           : 00               0-24 | Public
  DEFINITION SIGNATURE
   DEFinetime    :
   CHANGETime    :
   CHANGEUsrid   :
   CHANGEAGEnt   :                  CSDApi | CSDBatch
   CHANGEAGRel   :




  MESSAGES: 2 SEVERE
                                         SYSID=CICS APPLID=CICSTS42

PF 1 HELP 2 COM 3 END         6 CRSR 7 SBH 8 SFH 9 MSG 10 SB 11 SF 12 CNCL
```

48

You will need to enter the mapset name, the CICS group and a description for the transaction. This information is listed in the design, except for the group name. In this textbook we are specifying USER01 as our group name. In your case, you'll be assigned a group name by your system administrator (it might be your user id by default). Let's enter our info as follows:

```
DEF mapset
 OVERTYPE TO MODIFY                                    CICS RELEASE = 0670
  CEDA  DEFine MApset(          )
   MApset        ==> EMPMMNU
   Group         ==> USER01
   DEScription   ==> EMPLOYEE SUPPORT MAIN MENU
   REsident      ==> No              No | Yes
   USAge         ==> Normal          Normal | Transient
   USElpacopy    ==> No              No | Yes
   Status        ==> Enabled         Enabled | Disabled
   RSl            : 00               0-24 | Public
  DEFINITION SIGNATURE
   DEFinetime     :
   CHANGETime     :
   CHANGEUsrid    :
   CHANGEAGEnt    :                  CSDApi | CSDBatch
   CHANGEAGRel    :




                                        SYSID=CICS APPLID=CICSTS42

 PF 1 HELP 2 COM 3 END          6 CRSR 7 SBH 8 SFH 9 MSG 10 SB 11 SF 12 CNCL
```

When you press ENTER, you will see this showing the define is successful:

```
 OVERTYPE TO MODIFY                                    CICS RELEASE = 0670
  CEDA  DEFine MApset( EMPMMNU  )
   MApset         : EMPMMNU
   Group          : USER01
   DEScription   ==> EMPLOYEE SUPPORT MAIN MENU
   REsident      ==> No              No | Yes
   USAge         ==> Normal          Normal | Transient
   USElpacopy    ==> No              No | Yes
   Status        ==> Enabled         Enabled | Disabled
   RSl            : 00               0-24 | Public
  DEFINITION SIGNATURE
   DEFinetime     : 11/27/18 04:09:56
   CHANGETime     : 11/27/18 04:09:56
   CHANGEUsrid    : USER01
   CHANGEAGEnt    : CSDApi           CSDApi | CSDBatch
   CHANGEAGRel    : 0670


                                        SYSID=CICS APPLID=CICSTS42
  DEFINE SUCCESSFUL                     TIME: 04.09.56  DATE: 11/27/18
 PF 1 HELP 2 COM 3 END          6 CRSR 7 SBH 8 SFH 9 MSG 10 SB 11 SF 12 CNCL
```

Installing a Mapset

After defining your mapset, you must install it. You can follow the menu prompts again, or you can simply enter the full command as follows:

```
INSTALL MAPSET(EMPMMNU) GROUP(USER01)
 OVERTYPE TO MODIFY
  CEDA  Install
   ATomservice  ==>
   Bundle       ==>
   CONnection   ==>
   CORbaserver  ==>
   DB2Conn      ==>
   DB2Entry     ==>
   DB2Tran      ==>
   DJar         ==>
   DOctemplate  ==>
   Enqmodel     ==>
   File         ==>
   Ipconn       ==>
   JOurnalmodel ==>
   JVmserver    ==>
   LIBrary      ==>
   LSrpool      ==>
 +  MApset      ==> EMPMMNU


                                       SYSID=CICS APPLID=CICSTS42
   INSTALL SUCCESSFUL                  TIME: 04.12.08  DATE: 11/27/18
 PF 1 HELP       3 END         6 CRSR 7 SBH 8 SFH 9 MSG 10 SB 11 SF 12 CNCL
```

Displaying a Map

Now you can check the format of the screen by displaying the map in CICS. You cannot use the transaction name itself to display the map because we have yet not defined the transaction or program. However you can use the CECI utility as follows.

```
     CECI SEND MAP(EMPMNU) MAPSET(EMPMMNU)
```

Now you'll see this screen. Press ENTER.

```
SEND MAP(EMPMNU) MAPSET(EMPMMNU)
 STATUS:  ABOUT TO EXECUTE COMMAND                          NAME=
  EXEC CICS  SENd Map( 'EMPMNU ' )
   << FROm() > < LEngth() > < DAtaonly > | MAPOnly >
   < MAPSet( 'EMPMMNU' ) >
   < FMhparm() >
   < Reqid() >
   < LDc() | < ACTpartn() > < Outpartn() > >
   < MSr() >
   < Cursor() >
   < Set() < MAPPingdev() > | PAging | Terminal < Wait > < LAst > >
   < PRint >
   < FREekb >
   < ALArm >
   < L40 | L64 | L80 | Honeom >
   < NLeom >
   < ERASE < DEfault | ALTernate > | ERASEAup >
   < ACCum >
   < FRSet >
 + < NOflush >


PF 1 HELP 2 HEX 3 END 4 EIB 5 VAR 6 USER 7 SBH 8 SFH 9 MSG 10 SB 11 SF
```

Now you'll see the formatted map:

```
EMPMMNU                    EMPLOYEE SUPPORT MENU                    EMNU

          ENTER THE NUMBER OF YOUR SELECTION,   THEN PRESS ENTER.


                    _   1. EMPLOYEE INQUIRY

                        2. EMPLOYEE ADD

                        3. EMPLOYEE CHANGE

                        4. EMPLOYEE DELETE




  F3 EXIT
```

Congratulations, you've just created, deployed and displayed a CICS component!

Compiling and Viewing the Rest of the Screens

We could move on to the programming part of our project by starting the program EMP-PGMNU. Instead, I suggest defining, installing and displaying all your screens first. Doing so means you can perform your programming tasks without getting sidetracked by an improperly formatted screen. Let's go ahead and do the remaining four screen definitions.

Note: on the data entry screens, we initialize or pre-fill the enterable fields with X's or with the format required (e.g., YYYY-MM-DD for dates) until the user fills in data or the program fills it in with data from a query. An example appears below.

```
EMPMINQ                    EMPLOYEE INQUIRY                         EMIN


      EMPLOYEE ->              ENTER EMPLOYEE ID, THEN PRESS ENTER

      EMPLOYEE ID    XXXX

      EMP LAST NAME  XXXX

      EMP FIRST NAME XXXX

      EMP SOCIAL SEC XXXXXXXXX

      EMP YEARS SRVC 00

      EMP LAST PROM  YYYY-MM-DD




    F2 INQ   F3 EXIT   F4 ADD   F5 CHG   F6 DEL
```

This technique is not required, but I recommend doing the field pre-fill during development because it helps to see the data format. For production you can remove the pre-fill characters unless your user base finds it worthwhile. Some do, some don't.

EMPMINQ

Here is the BMS code for the employee inquiry screen EMPMINQ. Be sure to use tag names that make sense. The maintenance programmers that follow you will appreciate that.

```
EMPMINQ  DFHMSD TYPE=&SYSPARM,MODE=INOUT,CTRL=(FREEKB,FRSET),        X
                LANG=COBOL,TIOAPFX=YES,                              X
                DSATTS=(COLOR,HILIGHT),                              X
                MAPATTS=(COLOR,HILIGHT),                             X
                STORAGE=AUTO
EMPINQ   DFHMDI SIZE=(24,80)
         DFHMDF POS=(01,1),LENGTH=07,COLOR=BLUE,                     X
                INITIAL='EMPMINQ'
         DFHMDF POS=(01,28),LENGTH=16,COLOR=BLUE,                    X
                INITIAL='EMPLOYEE INQUIRY'
TRANID   DFHMDF POS=(01,76),LENGTH=04,INITIAL='EMIN',COLOR=BLUE
         DFHMDF POS=(04,08),LENGTH=11,COLOR=BLUE,                    X
                INITIAL='EMPLOYEE ->'
EMPIN    DFHMDF POS=(04,23),LENGTH=04,ATTRB=(IC,UNPROT),             X
                COLOR=GREEN,HILIGHT=UNDERLINE
         DFHMDF POS=(04,28),LENGTH=01,ATTRB=ASKIP
         DFHMDF POS=(04,34),LENGTH=35,COLOR=BLUE,                    X
                INITIAL='ENTER EMPLOYEE ID, THEN PRESS ENTER'
         DFHMDF POS=(06,08),LENGTH=14,COLOR=BLUE,                    X
                INITIAL='EMPLOYEE ID   '
EMPNO    DFHMDF POS=(06,23),LENGTH=04,COLOR=GREEN,                   X
                INITIAL='XXXX'
         DFHMDF POS=(08,08),LENGTH=14,COLOR=BLUE,                    X
                INITIAL='EMP LAST NAME '
LNAME    DFHMDF POS=(08,23),LENGTH=30,COLOR=GREEN,                   X
                INITIAL='XXXX'
         DFHMDF POS=(10,08),LENGTH=14,COLOR=BLUE,                    X
                INITIAL='EMP FIRST NAME'
FNAME    DFHMDF POS=(10,23),LENGTH=20,COLOR=GREEN,                   X
                INITIAL='XXXX'
         DFHMDF POS=(12,08),LENGTH=14,COLOR=BLUE,                    X
                INITIAL='EMP SOCIAL SEC'
SOCSEC   DFHMDF POS=(12,23),LENGTH=09,COLOR=GREEN,                   X
                INITIAL='XXXXXXXXX'
         DFHMDF POS=(14,08),LENGTH=14,COLOR=BLUE,                    X
                INITIAL='EMP YEARS SRVC'
YRSSVC   DFHMDF POS=(14,23),LENGTH=02,COLOR=GREEN,                   X
                INITIAL='00'
         DFHMDF POS=(16,08),LENGTH=14,COLOR=BLUE,                    X
                INITIAL='EMP LAST PROM '
LSTPRM   DFHMDF POS=(16,23),LENGTH=10,COLOR=GREEN,                   X
                INITIAL='YYYY-MM-DD'
MESSAGE  DFHMDF POS=(23,02),LENGTH=67,COLOR=YELLOW
         DFHMDF POS=(24,02),LENGTH=43,ATTRB=PROT,COLOR=BLUE,         X
                INITIAL='F2 INQ   F3 EXIT   F4 ADD   F5 CHG   F6 DEL'
         DFHMSD TYPE=FINAL
         END
```

Go ahead and compile the mapset. Then define, install and display it. The CICS commands are as follows:

```
DEF MAPSET
 OVERTYPE TO MODIFY                                      CICS RELEASE = 0670
  CEDA   DEFine MApset(            )
   MApset       ==> EMPMINQ
   Group        ==> USER01
   DEScription  ==> EMPLOYEE INQUIRY
   REsident     ==> No                  No | Yes
   USAge        ==> Normal              Normal | Transient
   USElpacopy   ==> No                  No | Yes
   Status       ==> Enabled             Enabled | Disabled
   RSl          : 00                    0-24 | Public
  DEFINITION SIGNATURE
  DEFinetime    :
  CHANGETime    :
  CHANGEUsrid   :
  CHANGEAGEnt   :                       CSDApi | CSDBatch
  CHANGEAGRel   :




                                              SYSID=CICS APPLID=CICSTS42

PF 1 HELP 2 COM 3 END          6 CRSR 7 SBH 8 SFH 9 MSG 10 SB 11 SF 12 CNCL


 INSTALL MAPSET(EMPMINQ) GROUP(USER01)
  OVERTYPE TO MODIFY
   CEDA  Install
    ATomservice  ==>
    Bundle       ==>
    CONnection   ==>
    CORbaserver  ==>
    DB2Conn      ==>
    DB2Entry     ==>
    DB2Tran      ==>
    DJar         ==>
    DOctemplate  ==>
    Enqmodel     ==>
    File         ==>
    Ipconn       ==>
    JOurnalmodel ==>
    JVmserver    ==>
    LIBrary      ==>
    LSrpool      ==>
 +  MApset       ==> EMPMINQ


                                              SYSID=CICS APPLID=CICSTS42
   INSTALL SUCCESSFUL                  TIME: 04.36.21  DATE: 11/27/18
PF 1 HELP        3 END          6 CRSR 7 SBH 8 SFH 9 MSG 10 SB 11 SF 12 CNCL
```

Now let's display the EMPINQ map using CECI:

```
CECI SEND MAP(EMPINQ) MAPSET(EMPMINQ)

SEND MAP(EMPINQ) MAPSET(EMPMINQ)
 STATUS:  ABOUT TO EXECUTE COMMAND                          NAME=
  EXEC CICS  SENd Map( 'EMPINQ ' )
   << FROm() > < LEngth() > < DAtaonly > | MAPOnly >
   < MAPSet( 'EMPMINQ' ) >
   < FMhparm() >
   < Reqid() >
   < LDc() | < ACTpartn() > < Outpartn() > >
   < MSr() >
   < Cursor() >
   < Set() < MAPPingdev() > | PAging | Terminal < Wait > < LAst > >
   < PRint >
   < FREekb >
   < ALArm >
   < L40 | L64 | L80 | Honeom >
   < NLeom >
   < ERASE < DEfault | ALTernate > | ERASEAup >
   < ACCum >
   < FRSet >
+  < NOflush >




 PF 1 HELP 2 HEX 3 END 4 EIB 5 VAR 6 USER 7 SBH 8 SFH 9 MSG 10 SB 11 SF
```

When you press ENTER you will see the Employee Inquiry screen map as follows.

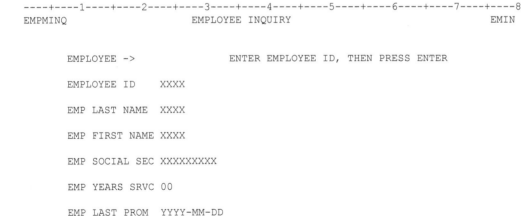

```
----+----1----+----2----+----3----+----4----+----5----+----6----+----7----+----8
EMPMINQ                    EMPLOYEE INQUIRY                              EMIN


       EMPLOYEE ->              ENTER EMPLOYEE ID, THEN PRESS ENTER

       EMPLOYEE ID    XXXX

       EMP LAST NAME  XXXX

       EMP FIRST NAME XXXX

       EMP SOCIAL SEC XXXXXXXXX

       EMP YEARS SRVC 00

       EMP LAST PROM  YYYY-MM-DD




   F2 INQ   F3 EXIT   F4 ADD   F5 CHG   F6 DEL
```

Now that you've got the hang of it, go ahead and compile, define, install and display the three other maps and mapsets. Makes sure they are as you want them to be displayed. I recommend that you try this on your own. You have the basics that you need, so give it a shot.

When you finish, turn the page and I'll give you my version of the three other mapsets. If you have any problems with compiling or displaying the maps, check to make sure your offsets are correct and that you don't have any overlapping fields.

EMPMADD

Here's the BMS code for the EMPMADD mapset.

```
EMPMADD  DFHMSD TYPE=&SYSPARM,MODE=INOUT,CTRL=(FREEKB,FRSET),          X
                LANG=COBOL,TIOAPFX=YES,                                X
                DSATTS=(COLOR,HILIGHT),                                X
                MAPATTS=(COLOR,HILIGHT),                               X
                STORAGE=AUTO
EMPADD   DFHMDI SIZE=(24,80)
         DFHMDF POS=(01,1),LENGTH=07,COLOR=BLUE,                       X
                INITIAL='EMPMADD'
         DFHMDF POS=(01,28),LENGTH=12,COLOR=BLUE,                      X
                INITIAL='EMPLOYEE ADD'
TRANID   DFHMDF POS=(01,76),LENGTH=04,INITIAL='EMAD',COLOR=BLUE
         DFHMDF POS=(04,23),LENGTH=35,COLOR=BLUE,                      X
                INITIAL='ENTER EMPLOYEE INFO, THEN PRESS PF4'
         DFHMDF POS=(06,08),LENGTH=14,COLOR=BLUE,                      X
                INITIAL='EMPLOYEE ID   '
EMPNO    DFHMDF POS=(06,23),LENGTH=04,COLOR=GREEN,                     X
                HILIGHT=UNDERLINE,ATTRB=(IC,UNPROT)
         DFHMDF POS=(06,28),LENGTH=01,ATTRB=ASKIP
         DFHMDF POS=(08,08),LENGTH=14,COLOR=BLUE,                      X
                INITIAL='EMP LAST NAME '
LNAME    DFHMDF POS=(08,23),LENGTH=30,COLOR=GREEN,ATTRB=(UNPROT),      X
                HILIGHT=UNDERLINE
         DFHMDF POS=(08,54),LENGTH=01,ATTRB=ASKIP
         DFHMDF POS=(10,08),LENGTH=14,COLOR=BLUE,                      X
                INITIAL='EMP FIRST NAME'
FNAME    DFHMDF POS=(10,23),LENGTH=20,COLOR=GREEN,ATTRB=(UNPROT),      X
                HILIGHT=UNDERLINE
         DFHMDF POS=(10,44),LENGTH=01,ATTRB=ASKIP
         DFHMDF POS=(12,08),LENGTH=14,COLOR=BLUE,                      X
                INITIAL='EMP SOCIAL SEC'
SOCSEC   DFHMDF POS=(12,23),LENGTH=09,COLOR=GREEN,ATTRB=(UNPROT),      X
                HILIGHT=UNDERLINE
         DFHMDF POS=(12,33),LENGTH=01,ATTRB=ASKIP
         DFHMDF POS=(14,08),LENGTH=14,COLOR=BLUE,                      X
                INITIAL='EMP YEARS SRVC'
YRSSVC   DFHMDF POS=(14,23),LENGTH=02,COLOR=GREEN,ATTRB=(UNPROT),      X
                HILIGHT=UNDERLINE
         DFHMDF POS=(14,26),LENGTH=01,ATTRB=ASKIP
         DFHMDF POS=(16,08),LENGTH=14,COLOR=BLUE,                      X
                INITIAL='EMP LAST PROM '
```

```
LSTPRM    DFHMDF POS=(16,23),LENGTH=10,COLOR=GREEN,ATTRB=(UNPROT),    X
              HILIGHT=UNDERLINE    `
          DFHMDF POS=(16,34),LENGTH=01,ATTRB=ASKIP
MESSAGE   DFHMDF POS=(23,02),LENGTH=67,COLOR=YELLOW
          DFHMDF POS=(24,02),LENGTH=43,ATTRB=PROT,COLOR=BLUE,         X
              INITIAL='F2 INQ   F3 EXIT   F4 ADD   F5 CHG   F6 DEL'
          DFHMSD TYPE=FINAL
          END
```

After you compile and install it, the add screen should display as follows using CECI:

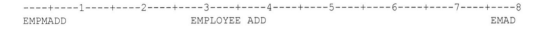

```
----+----1----+----2----+----3----+----4----+----5----+----6----+----7----+----8
EMPMADD                      EMPLOYEE ADD                             EMAD


        EMPLOYEE ->    ENTER EMPLOYEE INFO, THEN PRESS PF4PRESS ENTER

        EMPLOYEE ID

        EMP LAST NAME

        EMP FIRST NAME

        EMP SOCIAL SEC

        EMP YEARS SRVC

        EMP LAST PROM




   F2 INQ    F3 EXIT    F4 ADD    F5 CHG    F6 DEL
```

EMPMCHG

Now here's the BMS code for the change screen:

```
EMPMCHG   DFHMSD TYPE=&SYSPARM,MODE=INOUT,CTRL=(FREEKB,FRSET),        X
              LANG=COBOL,TIOAPFX=YES,                                 X
              DSATTS=(COLOR,HILIGHT),                                 X
              MAPATTS=(COLOR,HILIGHT),                                X
              STORAGE=AUTO
EMPCHG    DFHMDI SIZE=(24,80)
          DFHMDF POS=(01,1),LENGTH=07,COLOR=BLUE,                     X
              INITIAL='EMPMCHG'
          DFHMDF POS=(01,28),LENGTH=15,COLOR=BLUE,                    X
              INITIAL='EMPLOYEE CHANGE'
TRANID    DFHMDF POS=(01,76),LENGTH=04,INITIAL='EMCH',COLOR=BLUE
          DFHMDF POS=(04,08),LENGTH=11,COLOR=BLUE,                    X
              INITIAL='EMPLOYEE ->'
EMPIN     DFHMDF POS=(04,23),LENGTH=04,ATTRB=(IC,UNPROT),COLOR=GREEN, X
              HILIGHT=UNDERLINE
          DFHMDF POS=(04,28),LENGTH=01,ATTRB=ASKIP
          DFHMDF POS=(04,34),LENGTH=35,COLOR=BLUE,                    X
```

57

```
                 INITIAL='ENTER EMPLOYEE ID, THEN PRESS ENTER'
              DFHMDF POS=(06,08),LENGTH=14,COLOR=BLUE,              X
                 INITIAL='EMPLOYEE ID  '
EMPNO         DFHMDF POS=(06,23),LENGTH=04,COLOR=GREEN,             X
                 INITIAL='XXXX'
              DFHMDF POS=(08,08),LENGTH=14,COLOR=BLUE,              X
                 INITIAL='EMP LAST NAME '
LNAME         DFHMDF POS=(08,23),LENGTH=30,COLOR=GREEN,ATTRB=UNPROT, X
                 INITIAL='XXXX'
              DFHMDF POS=(10,08),LENGTH=14,COLOR=BLUE,              X
                 INITIAL='EMP FIRST NAME'
FNAME         DFHMDF POS=(10,23),LENGTH=20,COLOR=GREEN,ATTRB=UNPROT, X
                 INITIAL='XXXX'
              DFHMDF POS=(12,08),LENGTH=14,COLOR=BLUE,              X
                 INITIAL='EMP SOCIAL SEC'
SOCSEC        DFHMDF POS=(12,23),LENGTH=09,COLOR=GREEN,ATTRB=UNPROT, X
                 INITIAL='XXXXXXXXX'
              DFHMDF POS=(14,08),LENGTH=14,COLOR=BLUE,              X
                 INITIAL='EMP YEARS SRVC'
YRSSVC        DFHMDF POS=(14,23),LENGTH=02,COLOR=GREEN,ATTRB=UNPROT, X
                 INITIAL='00'
              DFHMDF POS=(16,08),LENGTH=14,COLOR=BLUE,              X
                 INITIAL='EMP LAST PROM '
LSTPRM        DFHMDF POS=(16,23),LENGTH=10,COLOR=GREEN,ATTRB=UNPROT, X
                 INITIAL='YYYY-MM-DD'
MESSAGE       DFHMDF POS=(23,02),LENGTH=67,COLOR=YELLOW
              DFHMDF POS=(24,02),LENGTH=45,ATTRB=PROT,COLOR=BLUE,   X
                 INITIAL='F2 INQ   F3 EXIT   F4 ADD   F5 CHG   F6 DEL'
              DFHMSD TYPE=FINAL
              END
```

The change screen should display as follows:

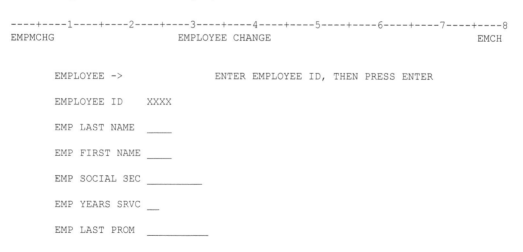

```
----+----1----+----2----+----3----+----4----+----5----+----6----+----7----+----8
EMPMCHG                     EMPLOYEE CHANGE                          EMCH


       EMPLOYEE ->              ENTER EMPLOYEE ID, THEN PRESS ENTER

       EMPLOYEE ID   XXXX

       EMP LAST NAME   ____

       EMP FIRST NAME ____

       EMP SOCIAL 3EC _____

       EMP YEARS SRVC __

       EMP LAST PROM   _____




  F2 INQ   F3 EXIT   F4 ADD   F5 CHG   F6 DEL
```

EMPMDEL

Now let's look at the BMS code for the delete screen.

```
EMPMDEL
EMPMDEL   DFHMSD TYPE=&SYSPARM,MODE=INOUT,CTRL=(FREEKB,FRSET),           X
              LANG=COBOL,TIOAPFX=YES,                                    X
              DSATTS=(COLOR,HILIGHT),                                    X
              MAPATTS=(COLOR,HILIGHT),                                   X
              STORAGE=AUTO
EMPDEL    DFHMDI SIZE=(24,80)
          DFHMDF POS=(01,1),LENGTH=07,COLOR=BLUE,                       X
              INITIAL='EMPMDEL'
          DFHMDF POS=(01,28),LENGTH=15,COLOR=BLUE,                      X
              INITIAL='EMPLOYEE DELETE'
TRANID    DFHMDF POS=(01,76),LENGTH=04,INITIAL='EMDE',COLOR=BLUE
          DFHMDF POS=(04,08),LENGTH=11,COLOR=BLUE,                      X
              INITIAL='EMPLOYEE ->'
EMPIN     DFHMDF POS=(04,23),LENGTH=04,ATTRB=(IC,UNPROT),COLOR=GREEN,   X
              HILIGHT=UNDERLINE
          DFHMDF POS=(04,28),LENGTH=01,ATTRB=ASKIP
          DFHMDF POS=(04,34),LENGTH=35,COLOR=BLUE,                      X
              INITIAL='ENTER EMPLOYEE ID, THEN PRESS ENTER'
          DFHMDF POS=(06,08),LENGTH=14,COLOR=BLUE,                      X
              INITIAL='EMPLOYEE ID   '
EMPNO     DFHMDF POS=(06,23),LENGTH=04,COLOR=GREEN,                     X
              INITIAL='XXXX'
          DFHMDF POS=(08,08),LENGTH=14,COLOR=BLUE,                      X
              INITIAL='EMP LAST NAME '
LNAME     DFHMDF POS=(08,23),LENGTH=30,COLOR=GREEN,ATTRB=UNPROT,        X
              INITIAL='XXXX'
          DFHMDF POS=(10,08),LENGTH=14,COLOR=BLUE,                      X
              INITIAL='EMP FIRST NAME'
FNAME     DFHMDF POS=(10,23),LENGTH=20,COLOR=GREEN,ATTRB=UNPROT,        X
              INITIAL='XXXX'
          DFHMDF POS=(12,08),LENGTH=14,COLOR=BLUE,                      X
              INITIAL='EMP SOCIAL SEC'
SOCSEC    DFHMDF POS=(12,23),LENGTH=09,COLOR=GREEN,ATTRB=UNPROT,        X
              INITIAL='XXXXXXXXX'
          DFHMDF POS=(14,08),LENGTH=14,COLOR=BLUE,                      X
              INITIAL='EMP YEARS SRVC'
YRSSVC    DFHMDF POS=(14,23),LENGTH=02,COLOR=GREEN,ATTRB=UNPROT,        X
              INITIAL='00'
          DFHMDF POS=(16,08),LENGTH=14,COLOR=BLUE,                      X
              INITIAL='EMP LAST PROM '
LSTPRM    DFHMDF POS=(16,23),LENGTH=10,COLOR=GREEN,ATTRB=UNPROT,        X
              INITIAL='YYYY-MM-DD'
MESSAGE   DFHMDF POS=(23,02),LENGTH=67,COLOR=YELLOW
          DFHMDF POS=(24,02),LENGTH=45,ATTRB=PROT,COLOR=BLUE,           X
              INITIAL='F2 INQ   F3 EXIT   F4 ADD   F5 CHG   F6 DEL'
          DFHMSD TYPE=FINAL
          END
```

The delete screen should display as follows:

```
----+----1----+----2----+----3----+----4----+----5----+----6----+----7----+----8
EMPMDEL                    EMPLOYEE DELETE                              EMDE


        EMPLOYEE ->               ENTER EMPLOYEE ID, THEN PRESS ENTER

        EMPLOYEE ID

        EMP LAST NAME

        EMP FIRST NAME

        EMP SOCIAL SEC

        EMP YEARS SRVC

        EMP LAST PROM




   F2 INQ   F3 EXIT   F4 ADD   F5 CHG   F6 DEL
```

Final Notes on Mapsets

When you make a modification to a mapset that has already been defined and installed into production, you must refresh the copy that resides in the online executable library. The reason is that a mapset is an executable program, and the load library that you compile into is not the same as the live load library that CICS uses to load programs. To refresh a modified program in CICS, you use the CEMT utility with the SET PROGRAM command and NEWCOPY parameter. For example, to refresh the menu program you would logon to CICS and enter the following:

```
CEMT SET PROGRAM(EMPMMNU) NEWCOPY
```

The above is also true for the application programs we'll soon be writing. Anytime you change the program, you must reload it into CICS using the CEMT utility.

Program Construction and Testing

Now we are ready to write the program code according to some basic specifications. For each program we will specify the name, purpose, data access requirements, SQL (if any), and some test cases. The test cases especially will help us understand how to code the program. Finally, we'll do unit testing of each program.

Employee Menu Program

The following are rudimentary specifications for our menu program.

Name: EMPPGMNU
Purpose: Employee Support Main Menu
Data Access: None
SQL: None (non-DB2)

EMPPGMNU Test Cases:

Note: four of the test cases below are shaded to indicate we cannot test them yet because they transfer to programs that we haven't written yet. We'll come back to those in integration testing.

Case	Condition	Expected Result	Actual Result
1	Initial screen display	All literals displayed, empty action field	
2	Invalid selection entered	Error message returned: ENTER A VALID ACTION: 1, 2, 3 OR 4	
3	Invalid key pressed	Error message: INVALID KEY PRESSED	
4	Option 1 entered	Transfer to EMIN transaction	
5	Option 2 entered	Transfer to EMAD transaction	
6	Option 3 entered	Transfer to EMCH transaction	
7	Option 4 entered	Transfer to EMDE transaction	

EMPPGMNU Program Code

We've already provided this code earlier as a demonstrator, but here it is again.

```
IDENTIFICATION DIVISION.
PROGRAM-ID. EMPPGMNU.
*****************************************************
*   MENU PROGRAM FOR EMPLOYEE APPLICATION          *
*                                                  *
*   AUTHOR        : ROBERT WINGATE                  *
*   DATE-WRITTEN  : 2018-07-26                      *
*****************************************************
ENVIRONMENT DIVISION.
```

```
DATA DIVISION.
WORKING-STORAGE SECTION.
01 WS-FLAGS.
    05 SW-VALID-SELECTION     PIC X(1) VALUE 'N'.
        88  VALID-SELECTION             VALUE 'Y'.
        88  NOT-VALID-SELECTION         VALUE 'N'.

01 WS-VARS.
    05 COMM-AREA              PIC X(20) VALUE SPACE.
    05 PROGRAM-NAME           PIC X(08) VALUE SPACES.
    05 INVALID-ACTION-MSG     PIC X(34)
        VALUE 'ENTER A VALID ACTION: 1, 2, 3 OR 4'.

    COPY EMPMMNU.
    COPY DFHAID.
    COPY DFHBMSCA.


LINKAGE SECTION.
01 DFHCOMMAREA          PIC X(20).


PROCEDURE DIVISION.

    IF EIBCALEN > ZERO
       MOVE DFHCOMMAREA  TO COMM-AREA
    END-IF.


    EVALUATE TRUE

       WHEN EIBCALEN = ZERO
          MOVE LOW-VALUES    TO  EMPMNUO
          PERFORM SEND-MAP


       WHEN EIBAID = DFHCLEAR
          MOVE LOW-VALUES    TO  EMPMNUO
          PERFORM SEND-MAP

       WHEN EIBAID = DFHPA1 OR DFHPA2 OR DFHPA3
          CONTINUE

       WHEN EIBAID = DFHPF3
          MOVE LOW-VALUES TO  EMPMNUO
          MOVE "BYE, PRESS CLEAR KEY TO ENTER A TRANSACTION ID"
               TO MESSAGEO
          PERFORM SEND-MAP-DATAONLY

          EXEC CICS
            RETURN
          END-EXEC


       WHEN EIBAID = DFHENTER
          PERFORM MAIN-PROCESS-PARA

       WHEN OTHER
          MOVE LOW-VALUES TO EMPMNUO
          MOVE "INVALID KEY PRESSED" TO MESSAGEO
```

```
            PERFORM SEND-MAP-DATAONLY

        END-EVALUATE.

        EXEC CICS
            RETURN TRANSID('EMNU')
            COMMAREA (COMM-AREA)
        END-EXEC.

    MAIN-PROCESS-PARA.

        PERFORM RECEIVE-MAP.
        IF ACTIONI NOT = '1' AND '2' AND '3' AND '4'
            MOVE DFHREVRS TO ACTIONH
            MOVE INVALID-ACTION-MSG TO MESSAGEO
            SET NOT-VALID-SELECTION TO TRUE
        ELSE
            SET VALID-SELECTION TO TRUE
        END-IF.

        IF VALID-SELECTION
            EVALUATE ACTIONI
                WHEN '1' MOVE 'EMPPGINQ' TO PROGRAM-NAME
                WHEN '2' MOVE 'EMPPGADD' TO PROGRAM-NAME
                WHEN '3' MOVE 'EMPPGCHG' TO PROGRAM-NAME
                WHEN '4' MOVE 'EMPPGDEL' TO PROGRAM-NAME
            END-EVALUATE

            PERFORM BRANCH-TO-PROGRAM

        END-IF.

        PERFORM SEND-MAP-DATAONLY.

    BRANCH-TO-PROGRAM.
        EXEC CICS
            XCTL PROGRAM(PROGRAM-NAME)
        END-EXEC

        MOVE 'PROGRAM NOT AVAILABLE' TO MESSAGEO.

    SEND-MAP.
        EXEC CICS SEND
            MAP    ('EMPMNU')
            MAPSET ('EMPMMNU')
            FROM   (EMPMNUO)
            ERASE
        END-EXEC.

    SEND-MAP-DATAONLY.
        EXEC CICS SEND
            MAP    ('EMPMNU')
            MAPSET ('EMPMMNU')
            FROM   (EMPMNUO)
            DATAONLY
        END-EXEC.
```

```
        RECEIVE-MAP.
            EXEC CICS RECEIVE
                MAP    ('EMPMNU')
                MAPSET ('EMPMMNU')
                INTO   (EMPMNUI)
            END-EXEC.
```

Compiling a Program

To compile the COBOL program, you'll need to locate the appropriate compile JCL used in your shop. Check with your supervisor or technical leader for the needed JCL. When you complete the compile process, always check the output for errors. Once there are no errors, you can proceed to define and install the program.

Here's the compile JCL I use:

```
//USER01D JOB MSGLEVEL=(1,1),NOTIFY=&SYSUID
//*  COBOL + DB2 + CICS COMPILE JCL
//*
//DBONL    EXEC DB2CICSC,
//              COPYLIB=USER01.COPYLIB,          <= COPYBOOK LIBRARY
//              DCLGLIB=USER01.DCLGEN.COBOL,     <= DCLGEN LIBRARY
//              DBRMLIB=USER01.DBRMLIB,          <= DBRM LIBRARY
//              SRCLIB=USER01.CICS.SRCLIB,       <= SOURCE LIBRARY
//              MEMBER=EMPPGADD                  <= SOURCE MEMBER
```

Defining a Program

Type DEF or DEFINE and press enter. You will see this screen.

```
DEF
 ENTER ONE OF THE FOLLOWING

Atomservice  MQconn       Webservice
Bundle       PARTItionset
CONnection   PARTNer
CORbaserver  PIpeline
DB2Conn      PROCesstype
DB2Entry     PROFile
DB2Tran      PROGram
DJar         Requestmodel
DOctemplate  Sessions
Enqmodel     TCpipservice
File         TDqueue
Ipconn       TErminal
JOurnalmodel TRANClass
JVmserver    TRANSaction
LIbrary      TSmodel
LSrpool      TYpeterm
MApset       Urimap


                                    SYSID=CICS APPLID=CICSTS42

PF 1 HELP      3 END        6 CRSR        9 MSG         12 CNCL
```

64

Now type PROGRAM and you will see this screen:

```
DEF PROGRAM
  OVERTYPE TO MODIFY                                    CICS RELEASE = 0670
   CEDA  DEFine PROGram(          )
    PROGram      ==>
    Group        ==>
    DEScription  ==>
    Language     ==>            CObol | Assembler | Le370 | C | Pli
    RELoad       ==> No         No | Yes
    RESident     ==> No         No | Yes
    USAge        ==> Normal     Normal | Transient
    USElpacopy   ==> No         No | Yes
    Status       ==> Enabled    Enabled | Disabled
    RSl          : 00           0-24 | Public
    CEdf         ==> Yes        Yes | No
    DAtalocation ==> Below      Below | Any
    EXECKey      ==> User       User | Cics
    COncurrency  ==> Quasirent  Quasirent | Threadsafe | Required
    Api          ==> Cicsapi    Cicsapi | Openapi
   REMOTE ATTRIBUTES
 +  DYnamic      ==> No         No | Yes
    MESSAGES: 2 SEVERE
                                        SYSID=CICS APPLID=CICSTS42

PF 1 HELP 2 COM 3 END          6 CRSR 7 SBH 8 SFH 9 MSG 10 SB 11 SF 12 CNCL
```

You will need to enter the mapset name, the CICS group and a description for the transaction. This information is in the design. So we enter it as follows:

```
DEF PROGRAM
  OVERTYPE TO MODIFY                                    CICS RELEASE = 0670
   CEDA  DEFine PROGram(          )
    PROGram      ==> EMPPGMNU
    Group        ==> USER01
    DEScription  ==> EMPLOYEE SUPPORT MAIN MENU
    Language     ==> COBOL      CObol | Assembler | Le370 | C | Pli
    RELoad       ==> No         No | Yes
    RESident     ==> No         No | Yes
    USAge        ==> Normal     Normal | Transient
    USElpacopy   ==> No         No | Yes
    Status       ==> Enabled    Enabled | Disabled
    RSl          : 00           0-24 | Public
    CEdf         ==> Yes        Yes | No
    DAtalocation ==> Below      Below | Any
    EXECKey      ==> User       User | Cics
    COncurrency  ==> Quasirent  Quasirent | Threadsafe | Required
    Api          ==> Cicsapi    Cicsapi | Openapi
   REMOTE ATTRIBUTES
 +  DYnamic      ==> No         No | Yes
    MESSAGES: 2 SEVERE
                                        SYSID=CICS APPLID=CICSTS42

PF 1 HELP 2 COM 3 END          6 CRSR 7 SBH 8 SFH 9 MSG 10 SB 11 SF 12 CNCL
```

65

When you press ENTER, you will see this screen, and always check for the DEFINE SUCCESSFUL message at the bottom:

```
OVERTYPE TO MODIFY                                    CICS RELEASE = 0670
  CEDA  DEFine PROGram( EMPPGMNU )
   PROGram        : EMPPGMNU
   Group          : USER01
   DEScription  ==> EMPLOYEE SUPPORT MAIN MENU
   Language     ==> COBOL              CObol | Assembler | Le370 | C | Pli
   RELoad       ==> No                 No | Yes
   RESident     ==> No                 No | Yes
   USAge        ==> Normal             Normal | Transient
   USElpacopy   ==> No                 No | Yes
   Status       ==> Enabled            Enabled | Disabled
   RSl            : 00                 0-24 | Public
   CEdf         ==> Yes                Yes | No
   DAtalocation ==> Below              Below | Any
   EXECKey      ==> User               User | Cics
   COncurrency  ==> Quasirent          Quasirent | Threadsafe | Required
   Api          ==> Cicsapi            Cicsapi | Openapi
  REMOTE ATTRIBUTES
+  DYnamic      ==> No                 No | Yes


                                            SYSID=CICS APPLID=CICSTS42
   DEFINE SUCCESSFUL                        TIME: 06.27.01  DATE: 11/27/18
 PF 1 HELP 2 COM 3 END          6 CRSR 7 SBH 8 SFH 9 MSG 10 SB 11 SF 12 CNCL
```

Installing a Program

Now you must install the program. You can follow the menu prompts again, or you can simply enter the full command as follows:

```
INSTALL PROGRAM(EMPPGMNU) GROUP(USER01)
 OVERTYPE TO MODIFY
  CEDA  Install
   ATomservice   ==>
   Bundle        ==>
   CONnection    ==>
   CORbaserver   ==>
   DB2Conn       ==>
   DB2Entry      ==>
   DB2Tran       ==>
   DJar          ==>
   DOctemplate   ==>
   Enqmodel      ==>
   File          ==>
   Ipconn        ==>
   JOurnalmodel  ==>
   JVmserver     ==>
   LIBrary       ==>
   LSrpool       ==>
+  MApset        ==>
                                            SYSID=CICS APPLID=CICSTS42
   INSTALL SUCCESSFUL                       TIME: 06.30.42  DATE: 11/27/18
 PF 1 HELP       3 END          6 CRSR 7 SBH 8 SFH 9 MSG 10 SB 11 SF 12 CNCL
```

Defining a Transaction

Next you must define and install the corresponding transaction. You do this in a similar way, but there is no source code. Go through the menu system or simply type DEF TRANS.

```
DEF TRANSACTION
 OVERTYPE TO MODIFY                                      CICS RELEASE = 0670
  CEDA  DEFine TRANSaction(      )
   TRANSaction   ==> EMNU
   Group         ==> USER01
   DEScription   ==> EMPLOYEE SUPPORT MAIN MENU
   PROGram       ==> EMPPGMNU
   TWasize       ==> 00000            0-32767
   PROFile       ==> DFHCICST
   PArtitionset  ==>
   STAtus        ==> Enabled          Enabled | Disabled
   PRIMedsize     : 00000             0-65520
   TASKDATALoc   ==> Below            Below | Any
   TASKDATAKey   ==> User             User | Cics
   STOrageclear  ==> No               No | Yes
   RUnaway       ==> System           System | 0 | 500-2700000
   SHutdown      ==> Disabled         Disabled | Enabled
   ISolate       ==> Yes              Yes | No
   Brexit        ==>
+ REMOTE ATTRIBUTES

                                       SYSID=CICS APPLID=CICSTS42

 PF 1 HELP 2 COM 3 END        6 CRSR 7 SBH 8 SFH 9 MSG 10 SB 11 SF 12 CNCL
```

```
 OVERTYPE TO MODIFY                                      CICS RELEASE = 0670
  CEDA  DEFine TRANSaction( EMNU )
   TRANSaction    : EMNU
   Group          : USER01
   DEScription   ==> EMPLOYEE SUPPORT MAIN MENU
   PROGram       ==> EMPPGMNU
   TWasize       ==> 00000            0-32767
   PROFile       ==> DFHCICST
   PArtitionset  ==>
   STAtus        ==> Enabled          Enabled | Disabled
   PRIMedsize     : 00000             0-65520
   TASKDATALoc   ==> Below            Below | Any
   TASKDATAKey   ==> User             User | Cics
   STOrageclear  ==> No               No | Yes
   RUnaway       ==> System           System | 0 | 500-2700000
   SHutdown      ==> Disabled         Disabled | Enabled
   ISolate       ==> Yes              Yes | No
   Brexit        ==>
+ REMOTE ATTRIBUTES

                                       SYSID=CICS APPLID=CICSTS42
  DEFINE SUCCESSFUL                    TIME: 06.38.01  DATE: 11/27/18
 PF 1 HELP 2 COM 3 END        6 CRSR 7 SBH 8 SFH 9 MSG 10 SB 11 SF 12 CNCL
```

Installing a Transaction

Now you must install the transaction, just like with the mapset and program. Here is the command and the result:

```
INSTALL TRANSACTION(EMNU) GROUP(USER01)
 OVERTYPE TO MODIFY
  CEDA  Install
   ATomservice   ==>
   Bundle        ==>
   CONnection    ==>
   CORbaserver   ==>
   DB2Conn       ==>
   DB2Entry      ==>
   DB2Tran       ==>
   DJar          ==>
   DOctemplate   ==>
   Enqmodel      ==>
   File          ==>
   Ipconn        ==>
   JOurnalmodel  ==>
   JVmserver     ==>
   LIBrary       ==>
   LSrpool       ==>
 + MApset        ==>


                                    SYSID=CICS APPLID=CICSTS42
   INSTALL SUCCESSFUL                TIME: 06.40.01  DATE: 11/27/18
 PF 1 HELP        3 END        6 CRSR 7 SBH 8 SFH 9 MSG 10 SB 11 SF 12 CNCL
```

Testing a Transaction

Now you can test the transaction, program and mapset by simply invoking the transaction. Let's try it. With a clear screen, enter transaction EMNU. It should display correctly on your screen.

```
EMPMMNU                     EMPLOYEE SUPPORT MENU                          EMNU

         ENTER THE NUMBER OF YOUR SELECTION,   THEN PRESS ENTER.


                         1. EMPLOYEE INQUIRY

                         2. EMPLOYEE ADD

                         3. EMPLOYEE CHANGE

                         4. EMPLOYEE DELETE






    F3 EXIT
```

Great, it displays!   However, we cannot test it further until the other programs are built and developed (so that the menu options can actually transfer you to those programs). So for now we'll place our menu program on the shelf as we construct the four DB2-CICS programs to access and operate on the employee data.

CICS DB2 Programs

Employee Inquiry Program

The following are basic specifications for our inquiry program. Pair this with the screen specifications and we can code this program.

Name: EMPPGINQ
Purpose: Display Employee Information
Data Access: Read-Only on EMPLOYEE
SQL:

```
SELECT EMP_ID,
       EMP_LAST_NAME,
       EMP_FIRST_NAME,
       EMP_SSN,
       EMP_SERVICE_YEARS,
       EMP_PROMOTION_DATE
  INTO
       :EMP-ID,
       :EMP-LAST-NAME,
       :EMP-FIRST-NAME,
       :EMP-SSN,
       :EMP-SERVICE-YEARS,
       :EMP-PROMOTION-DATE
  FROM USER01.EMPLOYEE
  WHERE EMP_ID = :EMP-ID
```

EMPPGINQ Test Cases:

Case	Condition	Expected Result	Actual Result
1	Initial screen display	All literals displayed, empty employee id	
2	Valid employee number entered	Detail returned for employee	
3	Invalid employee number entered	Error message that employee does not exist	
4	Enter pressed without changing anything	No change	
5	PF2 pressed	Refresh screen display	
6	PF3 pressed	Message to clear screen and enter a transaction id	
7	Other PF Keys pressed (PF1, PF7, PF8)	Error message – invalid key	
8	Employee id passed from other screen	Detail returned for employee	
9	PF4 pressed	Transfer to EMAD transaction	
10	PF5 pressed	Transfer to EMCH transaction	
11	PF6 Pressed	Transfer to EMDE transaction	

Note: the shaded test cases cannot be tested yet because the programs do not exist. We will bring these back and apply them in integration testing.

EMPPGINQ Program Code

The following is the program code for EMPPGINQ. Take a good look at it, and then we'll discuss how/why we coded it this way.

```
IDENTIFICATION DIVISION.
PROGRAM-ID. EMPPGINQ.
*****************************************************
*   COBOL/CICS/DB2 PROGRAM TO DISPLAY AN EMPLOYEE *
*   AUTHOR       : ROBERT WINGATE                 *
*   DATE-WRITTEN : 2018-07-19                     *
*****************************************************
ENVIRONMENT DIVISION.
DATA DIVISION.
WORKING-STORAGE SECTION.
01 WS-EMPNO         PIC 9(4).
01 WS-EMP-SRV-YRS   PIC 9(2).
01 WS-SQLCODE       PIC S9(08).
01 WS-COMMAREA.
   05 WS-EMP-PASS   PIC 9(04) VALUE ZERO.
   05 WS-PGM-PASS   PIC X(08) VALUE SPACES.
   05 FILLER        PIC X(08).

01 PROGRAM-NAME     PIC X(08) VALUE SPACES.
01 SW-PASSED-DATA-SWITCH   PIC X(1) VALUE 'N'.
   88  SW-PASSED-DATA              VALUE 'Y'.
   88  SW-NO-PASSED-DATA           VALUE 'N'.

   COPY EMPMINQ.
   COPY DFHAID.
   COPY DFHBMSCA.

   EXEC SQL
     INCLUDE SQLCA
   END-EXEC.

   EXEC SQL
     INCLUDE EMPLOYEE
   END-EXEC.

LINKAGE SECTION.
01 DFHCOMMAREA          PIC X(20).

PROCEDURE DIVISION.

   SET SW-NO-PASSED-DATA TO TRUE

   IF EIBCALEN > ZERO
     MOVE DFHCOMMAREA  TO WS-COMMAREA
   END-IF.
```

```
EVALUATE TRUE

  WHEN EIBCALEN = ZERO
    MOVE LOW-VALUES   TO  EMPINQO
    PERFORM SEND-MAP


  WHEN EIBAID = DFHCLEAR
    MOVE LOW-VALUES    TO  EMPINQO
    PERFORM SEND-MAP

  WHEN EIBAID = DFHPA1 OR DFHPA2 OR DFHPA3
    CONTINUE


  WHEN EIBAID = DFHPF2
    IF WS-PGM-PASS NOT EQUAL "EMPPGINQ"
       SET SW-PASSED-DATA TO TRUE
    ELSE
       SET SW-NO-PASSED-DATA TO TRUE
    END-IF
    PERFORM PROCESS-PARA

  WHEN EIBAID = DFHPF3
    MOVE LOW-VALUES TO  EMPINQO
    MOVE "BYE, PRESS CLEAR KEY TO ENTER A TRANSACTION ID"
         TO MESSAGEO

    PERFORM SEND-MAP-DATA

    EXEC CICS
      RETURN
    END-EXEC

  WHEN EIBAID = DFHPF4
    MOVE 'EMPPGADD' TO PROGRAM-NAME
    PERFORM BRANCH-TO-PROGRAM

    EXEC CICS
      RETURN
    END-EXEC


  WHEN EIBAID = DFHPF5
    MOVE 'EMPPGCHG' TO PROGRAM-NAME
    PERFORM BRANCH-TO-PROGRAM

    EXEC CICS
      RETURN
    END-EXEC

  WHEN EIBAID = DFHPF6
    MOVE 'EMPPGDEL' TO PROGRAM-NAME
    PERFORM BRANCH-TO-PROGRAM

    EXEC CICS
      RETURN
    END-EXEC
```

```
           WHEN EIBAID = DFHENTER
             PERFORM PROCESS-PARA

           WHEN OTHER
             MOVE LOW-VALUES TO EMPINQO
             MOVE "INVALID KEY PRESSED" TO MESSAGEO
             PERFORM SEND-MAP-DATA

       END-EVALUATE.

       EXEC CICS
          RETURN TRANSID('EMIN')
          COMMAREA (WS-COMMAREA)
          LENGTH(20)
       END-EXEC.

   PROCESS-PARA.

       PERFORM RECEIVE-MAP.
       INITIALIZE DCLEMPLOYEE MESSAGEO

       IF SW-PASSED-DATA
          MOVE WS-EMP-PASS TO EMP-ID
       ELSE
          MOVE EMPINI     TO WS-EMPNO WS-EMP-PASS
          MOVE WS-EMPNO   TO EMP-ID
       END-IF

       EXEC SQL
          SELECT EMP_ID,
                 EMP_LAST_NAME,
                 EMP_FIRST_NAME,
                 EMP_SSN,
                 EMP_SERVICE_YEARS,
                 EMP_PROMOTION_DATE
             INTO
                 :EMP-ID,
                 :EMP-LAST-NAME,
                 :EMP-FIRST-NAME,
                 :EMP-SSN,
                 :EMP-SERVICE-YEARS,
                 :EMP-PROMOTION-DATE
             FROM USER01.EMPLOYEE
             WHERE EMP_ID = :EMP-ID
       END-EXEC.

       MOVE SQLCODE   TO  WS-SQLCODE.
       EVALUATE SQLCODE
         WHEN 0
           MOVE EMP-ID               TO  WS-EMPNO
           MOVE WS-EMPNO             TO  EMPNOO EMPINO
           MOVE EMP-LAST-NAME-TEXT  TO LNAMEO
           MOVE EMP-FIRST-NAME-TEXT TO FNAMEO
           MOVE EMP-SSN             TO SOCSECO
           MOVE EMP-SERVICE-YEARS   TO WS-EMP-SRV-YRS
           MOVE WS-EMP-SRV-YRS      TO YRSSVCO
           MOVE EMP-PROMOTION-DATE  TO LSTPRMO
```

```
        WHEN 100
            STRING "EMPLOYEE ID " DELIMITED BY SIZE
            WS-EMPNO DELIMITED BY SPACE
            " NOT FOUND" DELIMITED BY SIZE INTO MESSAGEO
            MOVE WS-EMPNO       TO EMPINO
            MOVE SPACES         TO EMPNOO
            MOVE SPACES         TO LNAMEO
            MOVE SPACES         TO FNAMEO
            MOVE SPACES         TO SOCSECO
            MOVE SPACES         TO YRSSVCO
            MOVE SPACES         TO LSTPRMO


         WHEN OTHER
            STRING "ERROR - SQL CODE: " DELIMITED BY SIZE
                   WS-SQLCODE    DELIMITED BY SIZE
              INTO MESSAGEO
        END-EVALUATE.

        MOVE DFHBMFSE    TO EMPINF
        MOVE -1 TO EMPINL
        MOVE "EMPPGINQ" TO WS-PGM-PASS
        IF SW-PASSED-DATA
           PERFORM SEND-MAP
        ELSE
           PERFORM SEND-MAP-DATA
        END-IF.

BRANCH-TO-PROGRAM.
        EXEC CICS
            XCTL PROGRAM(PROGRAM-NAME)
            COMMAREA (WS-COMMAREA)
            LENGTH(20)
        END-EXEC

        MOVE 'PROGRAM NOT AVAILABLE' TO MESSAGEO.

SEND-MAP.
        EXEC CICS SEND
            MAP    ('EMPINQ')
            MAPSET ('EMPMINQ')
            FROM   (EMPINQO)
            ERASE
        END-EXEC.

SEND-MAP-DATA.
        EXEC CICS SEND
            MAP    ('EMPINQ')
            MAPSET ('EMPMINQ')
            FROM   (EMPINQO)
            DATAONLY
        END-EXEC.

RECEIVE-MAP.
        EXEC CICS RECEIVE
            MAP    ('EMPINQ')
            MAPSET ('EMPMINQ')
            INTO   (EMPINQI)
        END-EXEC.
```

Ok, first let's take a look at the working storage section below. We have a work variable for employee number to convert it to numeric for our DB2 query. Similarly we have a work variable for employee service years. We've added a picture field for the DB2 SQLCODE.

Then we have a communication area for transferring data between programs. Note that the communication area WS-COMMAREA has two fields: the passed employee id and the name of the program that passed the data. This is important because if another program passes an employee id, then we will first retrieve the data and load the symbolic map. Then we'll display the initial screen which will already be populated with data for the employee number that was passed to the program. We'll set a switch that will let us know there is passed data – this will help us to determine certain logic branches later in our processing of the user request.

Finally notice we include two DB2 copybooks. These are for the SQLCA and for the EMPLOYEE DB2 table.

```
WORKING-STORAGE SECTION.
01 WS-EMPNO          PIC 9(4).
01 WS-EMP-SRV-YRS    PIC 9(2).
01 WS-SQLCODE        PIC S9(08).
01 WS-COMMAREA.
   05 WS-EMP-PASS    PIC 9(04) VALUE ZERO.
   05 WS-PGM-PASS    PIC X(08) VALUE SPACES.
   05 FILLER         PIC X(08).
01 PROGRAM-NAME      PIC X(08) VALUE SPACES.
01 SW-PASSED-DATA-SWITCH   PIC X(1) VALUE 'N'.
   88  SW-PASSED-DATA               VALUE 'Y'.
   88  SW-NO-PASSED-DATA            VALUE 'N'.

   COPY EMPMINQ.
   COPY DFHAID.
   COPY DFHBMSCA.

   EXEC SQL
     INCLUDE SQLCA
   END-EXEC.

   EXEC SQL
     INCLUDE EMPLOYEE
   END-EXEC.

LINKAGE SECTION.

01 DFHCOMMAREA        PIC X(20).
```

Now let's take a look at the procedures to send and receive data. These look just like the ones in the menu program, except for the names of the map and mapset. The initial screen will be sent using SEND-MAP. Subsequent screens will be sent using SEND-MAP-DATA. The user's screen input will be received into the program and loaded to the symbolic map using RECEIVE-MAP.

```
SEND-MAP.
    EXEC CICS SEND
        MAP     ('EMPINQ')
        MAPSET  ('EMPMINQ')
        FROM    (EMPINQO)
        ERASE
    END-EXEC.

SEND-MAP-DATA.
    EXEC CICS SEND
        MAP     ('EMPINQ')
        MAPSET  ('EMPMINQ')
        FROM    (EMPINQO)
        DATAONLY
    END-EXEC.


RECEIVE-MAP.
    EXEC CICS RECEIVE
        MAP     ('EMPINQ')
        MAPSET  ('EMPMINQ')
        INTO    (EMPINQI)
    END-EXEC.
```

Now let's look briefly at the BRANCH-TO-PROGRAM paragraph. This one is slightly different. Besides the program name to transfer to, we've added two parameters: COMMAREA and LENGTH. These are somewhat self-explanatory (the named communication area we defined in working storage, and it's length), and they are absolutely necessary to transfer data from one program to another. We're going to code for that but we won't be able to test that part of the program until the other programs are completed.

```
BRANCH-TO-PROGRAM.

    EXEC CICS
        XCTL PROGRAM(PROGRAM-NAME)
        COMMAREA (WS-COMMAREA)
        LENGTH(20)
    END-EXEC

    MOVE 'PROGRAM NOT AVAILABLE' TO MESSAGEO.
```

Now let's look at our program mainline and see what's different from the menu program. For our inquiry program, if the pressed key is PF2 and the name of the program passing data is not this program's name (EMPPGINQ), then we know this program was arrived at by transfer from another program. So we turn on the SW-PASSED-DATA switch. We'll use that later. In either case, since the PF2 key was pressed we call the processing paragraph. We'll look at that momentarily.

Meanwhile, notice that if the PF4, PF5 or PF6 key is pressed, the program will perform a transfer to the add, change or delete program, respectively. Since we haven't programmed those yet, you might want to comment out that code (or just remember not to press those keys in testing until the programs are written – otherwise your session will freeze or possibly abend).

```
            SET SW-NO-PASSED-DATA TO TRUE

            IF EIBCALEN > ZERO
              MOVE DFHCOMMAREA  TO WS-COMMAREA
            END-IF.

            EVALUATE TRUE

              WHEN EIBCALEN = ZERO
                MOVE LOW-VALUES   TO  EMPINQO
                PERFORM SEND-MAP

              WHEN EIBAID = DFHCLEAR
                MOVE LOW-VALUES    TO  EMPINQO
                PERFORM SEND-MAP

              WHEN EIBAID = DFHPA1 OR DFHPA2 OR DFHPA3
                CONTINUE

              WHEN EIBAID = DFHPF2
                IF WS-PGM-PASS NOT EQUAL "EMPPGINQ"
                   SET SW-PASSED-DATA TO TRUE
                ELSE
                   SET SW-NO-PASSED-DATA TO TRUE
                END-IF
                PERFORM PROCESS-PARA

              WHEN EIBAID = DFHPF3
                MOVE LOW-VALUES TO  EMPINQO
                MOVE "BYE, PRESS CLEAR KEY TO ENTER A TRANSACTION ID"
                     TO MESSAGEO
                PERFORM SEND-MAP-DATA

                EXEC CICS
                  RETURN
                END-EXEC

              WHEN EIBAID = DFHPF4
                MOVE 'EMPPGADD' TO PROGRAM-NAME
                PERFORM BRANCH-TO-PROGRAM

                EXEC CICS
                  RETURN
```

```
                END-EXEC

          WHEN EIBAID = DFHPF5
            MOVE 'EMPPGCHG' TO PROGRAM-NAME
            PERFORM BRANCH-TO-PROGRAM

            EXEC CICS
              RETURN
            END-EXEC

          WHEN EIBAID = DFHPF6
            MOVE 'EMPPGDEL' TO PROGRAM-NAME
            PERFORM BRANCH-TO-PROGRAM

            EXEC CICS
              RETURN
            END-EXEC

          WHEN EIBAID = DFHENTER
            PERFORM PROCESS-PARA

          WHEN OTHER
            MOVE LOW-VALUES TO EMPINQO
            MOVE "INVALID KEY PRESSED" TO MESSAGEO
            PERFORM SEND-MAP-DATA

        END-EVALUATE.

        EXEC CICS
           RETURN TRANSID('EMIN')
           COMMAREA (WS-COMMAREA)
           LENGTH(20)
        END-EXEC.
```

Now let's look at the main processing paragraph. Here we will load the employee id to be
retrieved, do the DB2 processing, and then either load the screen with employee data (if
successful) or report an error. Then we'll perform a send map action.

```
        PROCESS-PARA.

            PERFORM RECEIVE-MAP.
            INITIALIZE DCLEMPLOYEE MESSAGEO

            IF SW-PASSED-DATA
               MOVE WS-EMP-PASS TO EMP-ID
            ELSE
               MOVE EMPINI    TO WS-EMPNO WS-EMP-PASS
               MOVE WS-EMPNO  TO EMP-ID
            END-IF

            EXEC SQL
               SELECT EMP_ID,
                      EMP_LAST_NAME,
                      EMP_FIRST_NAME,
                      EMP_SSN,
                      EMP_SERVICE_YEARS,
```

```
               EMP_PROMOTION_DATE
        INTO
            :EMP-ID,
            :EMP-LAST-NAME,
            :EMP-FIRST-NAME,
            :EMP-SSN,
            :EMP-SERVICE-YEARS,
            :EMP-PROMOTION-DATE
        FROM USER01.EMPLOYEE
        WHERE EMP_ID = :EMP-ID
END-EXEC.

MOVE SQLCODE   TO   WS-SQLCODE.

EVALUATE SQLCODE
   WHEN 0
      MOVE EMP-ID                 TO  WS-EMPNO
      MOVE WS-EMPNO               TO  EMPNOO EMPINO
      MOVE EMP-LAST-NAME-TEXT   TO LNAMEO
      MOVE EMP-FIRST-NAME-TEXT  TO FNAMEO
      MOVE EMP-SSN               TO SOCSECO
      MOVE EMP-SERVICE-YEARS    TO WS-EMP-SRV-YRS
      MOVE WS-EMP-SRV-YRS       TO YRSSVCO
      MOVE EMP-PROMOTION-DATE   TO LSTPRMO

   WHEN 100
      STRING "EMPLOYEE ID " DELIMITED BY SIZE
      WS-EMPNO DELIMITED BY SPACE
      " NOT FOUND" DELIMITED BY SIZE INTO MESSAGEO
      MOVE WS-EMPNO      TO EMPINO
      MOVE SPACES        TO EMPNOO
      MOVE SPACES        TO LNAMEO
      MOVE SPACES        TO FNAMEO
      MOVE SPACES        TO SOCSECO
      MOVE SPACES        TO YRSSVCO
      MOVE SPACES        TO LSTPRMO

   WHEN OTHER
      STRING "ERROR - SQL CODE: " DELIMITED BY SIZE
             WS-SQLCODE    DELIMITED BY SIZE
        INTO MESSAGEO
END-EVALUATE.

MOVE DFHBMFSE    TO EMPINF
MOVE -1 TO EMPINL
MOVE "EMPPGINQ" TO WS-PGM-PASS

IF SW-PASSED-DATA
   PERFORM SEND-MAP
ELSE
   PERFORM SEND-MAP-DATA
END-IF.
```

Notice that we check SW-PASSED-DATA to see if data was passed to the program, and if it was, we use the passed employee id value for the query. Otherwise we will use the employee id the user entered on the screen which resides in the input symbolic map.

Also we load the inquiry program name into `WS-PGM-PASS` so on the next time through the program we know to use whatever employee id is on the screen (the user may have changed it) instead of the previously passed employee id.

Finally, we check to see if data was passed and if so we will use the send map routine that clears the screen and writes the entire map (including literals) rather than the data-only version that is used when the user has entered something on an already presented employee inquiry screen.

Now we can compile the program, define and install it in CICS, and test it as well. Again you will need to check with your supervisor or a fellow programmer for the correct JCL to compile a CICS-DB2 program. Mine looks like this:

```
//USER01D JOB MSGLEVEL=(1,1),NOTIFY=&SYSUID
//*
//*  COBOL + DB2 + CICS COMPILE JCL
//*
//DBONL    EXEC DB2CICSC,
//              COPYLIB=USER01.COPYLIB,          <= COPYBOOK LIBRARY
//              DCLGLIB=USER01.DCLGEN.COBOL,     <= DCLGEN LIBRARY
//              DBRMLIB=USER01.DBRMLIB,          <= DBRM LIBRARY
//              SRCLIB=USER01.CICS.SRCLIB,       <= SOURCE LIBRARY
//              MEMBER=EMPPGCHG                   <= SOURCE MEMBER
```

You will also need to bind the program of course. I use this DB2 command to do it:

```
BIND  MEMBER    (EMPPGINQ) -
PLAN      (MRWP01) -
ACTION    (REPLACE)  -
ISOLATION (CS)       -
EXPLAIN   (YES)      -
VALIDATE  (BIND)     -
RELEASE   (COMMIT)   -
OWNER     (USER01)   -
QUALIFIER (USER01)   -
ENCODING  (1047)
```

Take particular note of the plan name because you'll need it when defining the `DB2ENTRY` CICS entity that is required for each DB2 program. Note also that my plan name is rather arbitrary. You will need to use a plan name that is consistent with your installation's standards. Sometimes a DBA will assign the plan name for you to use.

Ok, next you can define the program and transaction exactly the way you defined the menu program:

```
OVERTYPE TO MODIFY                                       CICS RELEASE = 0670
  CEDA  DEFine PROGram( EMPPGINQ )
   PROGram       : EMPPGINQ
   Group         : USER01
   DEScription  ==> EMPLOYEE INQUIRY
   Language     ==> CObol              CObol | Assembler | Le370 | C | Pli
   RELoad       ==> No                 No | Yes
   RESident     ==> No                 No | Yes
   USAge        ==> Normal             Normal | Transient
   USElpacopy   ==> No                 No | Yes
   Status       ==> Enabled            Enabled | Disabled
   RSl           : 00                  0-24 | Public
   CEdf         ==> Yes                Yes | No
   DAtalocation ==> Below              Below | Any
   EXECKey      ==> User               User | Cics
   COncurrency  ==> Quasirent          Quasirent | Threadsafe | Required
   Api          ==> Cicsapi            Cicsapi | Openapi
  REMOTE ATTRIBUTES
+  DYnamic      ==> No                 No | Yes

                                          SYSID=CICS APPLID=CICSTS42
  DEFINE SUCCESSFUL                       TIME: 07.26.57  DATE: 11/27/18
PF 1 HELP 2 COM 3 END        6 CRSR 7 SBH 8 SFH 9 MSG 10 SB 11 SF 12 CNCL



OVERTYPE TO MODIFY                                       CICS RELEASE = 0670
  CEDA  DEFine TRANSaction( EMIN )
   TRANSaction   : EMIN
   Group         : USER01
   DEScription  ==> EMPLOYEE INQUIRY
   PROGram      ==> EMPPGINQ
   TWasize      ==> 00000              0-32767
   PROFile      ==> DFHCICST
   PArtitionset ==>
   STAtus       ==> Enabled            Enabled | Disabled
   PRIMedsize    : 00000               0-65520
   TASKDATALoc  ==> Below              Below | Any
   TASKDATAKey  ==> User               User | Cics
   STOrageclear ==> No                 No | Yes
   RUnaway      ==> System             System | 0 | 500-2700000
   SHutdown     ==> Disabled           Disabled | Enabled
   ISolate      ==> Yes                Yes | No
   Brexit       ==>
+ REMOTE ATTRIBUTES
                                          SYSID=CICS APPLID=CICSTS42
  DEFINE SUCCESSFUL                       TIME: 07.32.26  DATE: 11/27/18
PF 1 HELP 2 COM 3 END        6 CRSR 7 SBH 8 SFH 9 MSG 10 SB 11 SF 12 CNCL
```

And use these commands to install the program and transaction.

```
     CEDA INSTALL PROGRAM(EMPPGINQ) GROUP(USER01)

     CEDA INSTALL TRANS(EMIN) GROUP(USER01)
```

Finally, we need a `DB2ENTRY` to provide DB2 connectivity to the transaction. Here's how to define it – make sure to enter your DB2 plan name in the `PLAN` field and your transaction id in the transid field. Otherwise your program will fail when you run it.

```
 DEF DB2ENTRY
 OVERTYPE TO MODIFY                                        CICS RELEASE = 0670
  CEDA  DEFine DB2Entry(            )
   DB2Entry      ==> EMPPGINQ
   Group         ==> USER01
   DEScription   ==> EMPLOYEE INQUIRY
  THREAD SELECTION ATTRIBUTES
   TRansid       ==> EMIN
  THREAD OPERATION ATTRIBUTES
   ACcountrec    ==> None              None | TXid | TAsk | Uow
   AUTHId        ==>
   AUTHType      ==>                   Userid | Opid | Group | Sign | TErm
                                       | TX
   DRollback     ==> Yes               Yes | No
   PLAN          ==> MRWP01
   PLANExitname  ==>
   PRIority      ==> High              High | Equal | Low
   PROtectnum    ==> 0000              0-2000
   THREADLimit   ==>                   0-2000
 + THREADWait    ==> Pool              Pool | Yes | No


                                            SYSID=CICS APPLID=CICSTS42

 PF 1 HELP 2 COM 3 END           6 CRSR 7 SBH 8 SFH 9 MSG 10 SB 11 SF 12 CNCL
```

```
 OVERTYPE TO MODIFY                                        CICS RELEASE = 0670
  CEDA  DEFine DB2Entry( EMPPGINQ )
   DB2Entry      : EMPPGINQ
   Group         : USER01
   DEScription   ==> EMPLOYEE INQUIRY
  THREAD SELECTION ATTRIBUTES
   TRansid       ==> EMIN
  THREAD OPERATION ATTRIBUTES
   ACcountrec    ==> None              None | TXid | TAsk | Uow
   AUTHId        ==>
   AUTHType      ==> Userid            Userid | Opid | Group | Sign | TErm
                                       | TX
   DRollback     ==> Yes               Yes | No
   PLAN          ==> MRWP01
   PLANExitname  ==>
   PRIority      ==> High              High | Equal | Low
   PROtectnum    ==> 0000              0-2000
   THREADLimit   ==> 0000              0-2000
 + THREADWait    ==> Pool              Pool | Yes | No


                                            SYSID=CICS APPLID=CICSTS42
  DEFINE SUCCESSFUL                       TIME: 07.42.25  DATE: 11/27/18
 PF 1 HELP 2 COM 3 END           6 CRSR 7 SBH 8 SFH 9 MSG 10 SB 11 SF 12 CNCL
```

Now you can install the DB2ENTRY as follows:

```
CEDA INSTALL DB2ENTRY(EMPPGINQ) GROUP(USER01)
```

And now we are ready to test our inquiry program! Enter the EMIN transaction in CICS and you should see this screen.

```
EMPMINQ                    EMPLOYEE INQUIRY                          EMIN


     EMPLOYEE ->               ENTER EMPLOYEE ID, THEN PRESS ENTER

     EMPLOYEE ID    XXXX

     EMP LAST NAME  XXXX

     EMP FIRST NAME XXXX

     EMP SOCIAL SEC XXXXXXXXX

     EMP YEARS SRVC 00

     EMP LAST PROM  YYYY-MM-DD




  F2 INQ   F3 EXIT   F4 ADD   F5 CHG   F6 DEL
```

Now key employee id 3217 into the EMPLOYEE field and press ENTER. Your result should be similar to this:

```
EMPMINQ                    EMPLOYEE INQUIRY                          EMIN


     EMPLOYEE ->   3217       ENTER EMPLOYEE ID, THEN PRESS ENTER

     EMPLOYEE ID    3217

     EMP LAST NAME  JOHNSON

     EMP FIRST NAME EDWARD

     EMP SOCIAL SEC 397342007

     EMP YEARS SRVC 07

     EMP LAST PROM  2017-01-01




  F2 INQ   F3 EXIT   F4 ADD   F5 CHG   F6 DEL
```

There you go! You've completed a CICS DB2 screen, program, transaction and DB2Entry. Now go ahead and execute all the test cases except the ones that transfer to other programs (PF4, PF5 and PF6), or that process data that is received from these programs. Make sure all functions work as intended before going on.

Now we have three more programs to write. These will be very similar to the inquiry program except of course we will be adding, updating or deleting data. That means we'll need field edits and the use of attributes to place the cursor, and a few other new features. Let's proceed with the add program.

Employee Add Program

The following are specifications for our add program.

Name: EMPPGADD
Purpose: Add a new employee to the EMPLOYEE table.
Data Access: INSERT on EMPLOYEE
SQL Statements:

```
INSERT INTO USER01.EMPLOYEE
(EMP_ID,
 EMP_LAST_NAME,
 EMP_FIRST_NAME,
 EMP_SERVICE_YEARS,
 EMP_PROMOTION_DATE,
 EMP_SSN)

VALUES
(:EMP-ID,
 :EMP-LAST-NAME,
 :EMP-FIRST-NAME,
 :EMP-SERVICE-YEARS,
 :EMP-PROMOTION-DATE,
 :EMP-SSN)
```

EMPPGADD Test Cases:

Case	Condition	Expected Result	Actual Result
1	Initial screen display	Blank with enterable fields	
2	Enter pressed without changing anything	No change	
3	Employee number is blank	Error – employee number is required	
4	Employee number not numeric	Error – employee number must be numeric	
5	Last name is blank	Error – Last Name is required	
6	First name is blank	Error – Fast Name is required	

7	Social Security number is blank	Error – Social Security Number is required	
8	Social Security number is not numeric	Error – Social Security Number must be numeric	
9	Years of Service is blank	Error – Years of Service is required	
10	Years of Service not numeric	Error – Years of Service must be numeric	
11	Last Promotion Date is blank	Error – Last Promotion Date is required	
12	PF2 pressed	Transfer to EMIN transaction	
13	PF3 pressed	Message to clear screen and enter a transaction id	
14	PF4 pressed	Add record if no errors, message that record was added successfully	
15	PF5 pressed	Transfer to EMCH transaction	
16	PF6 Pressed	Transfer to EMDE transaction	
17	Other PF Keys pressed (PF1, PF7, PF8)	Error message – invalid key	

Employee Add Program Code

Here is the code for the add program. Much of the program looks like the inquiry program. But please review this add program carefully to see how we handle certain requirements such as field edits and errors. Also notice that the "action" key for this transaction is PF4 (not ENTER).

```
IDENTIFICATION DIVISION.
PROGRAM-ID. EMPPGADD.
***************************************************
*   COBOL/CICS/DB2 PROGRAM TO ADD AN EMPLOYEE     *
*                                                 *
*   AUTHOR        : ROBERT WINGATE                *
*   DATE-WRITTEN  : 2018-07-21                    *
***************************************************
ENVIRONMENT DIVISION.
DATA DIVISION.
WORKING-STORAGE SECTION.
01 WS-EMPNO         PIC 9(04).
01 WS-EMP-SRV-YRS   PIC 9(02).
01 WS-SQLCODE       PIC 9(08).
01 WS-COMMAREA.
   05 WS-EMP-PASS   PIC 9(04).
   05 WS-PGM-PASS   PIC X(08).
   05 FILLER        PIC X(08).

01 PROGRAM-NAME     PIC X(08) VALUE SPACES.
01 POS-CTR          PIC S9(9) USAGE COMP VALUE +0.

01  SW-SPACE-FOUND-SWITCH   PIC X(1) VALUE 'N'.
    88  SW-SPACE-FOUND              VALUE 'Y'.
    88  SW-SPACE-NOT-FOUND          VALUE 'N'.
```

```
        COPY EMPMADD.
        COPY DFHAID.
        COPY DFHBMSCA.

        EXEC SQL
          INCLUDE SQLCA
        END-EXEC.

        EXEC SQL
          INCLUDE EMPLOYEE
        END-EXEC.

LINKAGE SECTION.

01 DFHCOMMAREA          PIC X(20).

PROCEDURE DIVISION.

        IF EIBCALEN > ZERO
          MOVE DFHCOMMAREA  TO WS-COMMAREA
        END-IF.

        EVALUATE TRUE

          WHEN EIBCALEN = ZERO
            MOVE LOW-VALUES   TO  EMPADDO
            MOVE -1 TO EMPNOL
            PERFORM SEND-MAP

          WHEN EIBAID = DFHCLEAR
            MOVE LOW-VALUES   TO  EMPADDO
            MOVE -1 TO EMPNOL
            PERFORM SEND-MAP

          WHEN EIBAID = DFHPA1 OR DFHPA2 OR DFHPA3

            CONTINUE

          WHEN EIBAID = DFHPF2
            MOVE 'EMPPGINQ' TO PROGRAM-NAME
            PERFORM BRANCH-TO-PROGRAM

            EXEC CICS
              RETURN
            END-EXEC

          WHEN EIBAID = DFHPF3
            MOVE LOW-VALUES TO  EMPADDO
            MOVE -1 TO EMPNOL
            MOVE "BYE, PRESS CLEAR KEY TO ENTER A TRANSACTION ID"
                TO MESSAGEO
            PERFORM SEND-MAP-DATA

            EXEC CICS
              RETURN
            END-EXEC

          WHEN EIBAID = DFHPF4
```

86

```
*          PERFORM THE EDITS AND VALIDATIONS
*          IF NO ERRORS THEN INSERT THE RECORDS


       IF WS-PGM-PASS NOT EQUAL "EMPPGADD"
          MOVE LOW-VALUES TO EMPADDO
          MOVE -1 TO EMPNOL
          MOVE DFHBMFSE TO EMPNOF
          MOVE
          "ENTER DATA FOR NEW EMPLOYEE, THEN PRESS PF4 TO ADD"
             TO MESSAGEO
          PERFORM SEND-MAP
       ELSE
          PERFORM VALIDATE-DATA
       END-IF

    WHEN EIBAID = DFHPF5
      MOVE 'EMPPGCHG' TO PROGRAM-NAME
      PERFORM BRANCH-TO-PROGRAM

    WHEN EIBAID = DFHPF6
      MOVE 'EMPPGDEL' TO PROGRAM-NAME
      PERFORM BRANCH-TO-PROGRAM

    WHEN EIBAID = DFHENTER
      PERFORM PROCESS-PARA

    WHEN OTHER
      MOVE LOW-VALUES TO EMPADDO
      MOVE -1 TO EMPNOL
      MOVE "INVALID KEY PRESSED" TO MESSAGEO
      PERFORM SEND-MAP-DATA

  END-EVALUATE.

  EXEC CICS
     RETURN TRANSID('EMAD')
     COMMAREA (WS-COMMAREA)
  END-EXEC.

PROCESS-PARA.

  PERFORM RECEIVE-MAP.
  INITIALIZE DCLEMPLOYEE MESSAGEO

  MOVE DFHBMFSE TO EMPNOF
  MOVE DFHBMFSE TO LNAMEF
  MOVE DFHBMFSE TO FNAMEF
  MOVE DFHBMFSE TO SOCSECF
  MOVE DFHBMFSE TO YRSSVCF
  MOVE DFHBMFSE TO LSTPRMF

  MOVE -1 TO EMPNOL.
  MOVE "ENTER DATA FOR NEW EMPLOYEE, THEN PRESS PF4 TO ADD"
       TO MESSAGEO

  MOVE "EMPPGADD" TO WS-PGM-PASS
  PERFORM SEND-MAP-ALL.
```

```
VALIDATE-DATA.

    PERFORM RECEIVE-MAP
    INITIALIZE DCLEMPLOYEE MESSAGEO

    MOVE DFHBMFSE TO EMPNOF
    MOVE DFHBMFSE TO LNAMEF
    MOVE DFHBMFSE TO FNAMEF
    MOVE DFHBMFSE TO SOCSECF
    MOVE DFHBMFSE TO YRSSVCF
    MOVE DFHBMFSE TO LSTPRMF

    EVALUATE TRUE

        WHEN EMPNOI EQUAL SPACES OR EMPNOL EQUAL ZERO
            MOVE "EMPLOYEE NUMBER IS REQUIRED" TO MESSAGEO
            MOVE -1 TO EMPNOL

        WHEN EMPNOI IS NOT NUMERIC
            MOVE "EMPLOYEE NUMBER MUST BE NUMERIC" TO MESSAGEO
            MOVE -1 TO EMPNOL

        WHEN LNAMEI EQUAL SPACES OR LNAMEL EQUAL ZERO
            MOVE "EMPLOYEE LAST NAME IS REQUIRED" TO MESSAGEO
            MOVE -1 TO LNAMEL

        WHEN FNAMEI EQUAL SPACES OR FNAMEL EQUAL ZERO
            MOVE "EMPLOYEE FIRST NAME IS REQUIRED" TO MESSAGEO
            MOVE -1 TO FNAMEL

        WHEN SOCSECI EQUAL SPACES OR SOCSECL EQUAL ZERO
            MOVE "SOCIAL SECURITY NUMBER IS REQUIRED" TO MESSAGEO
            MOVE -1 TO SOCSECL

        WHEN SOCSECI IS NOT NUMERIC
            MOVE "SOCIAL SECURITY MUST BE NUMERIC" TO MESSAGEO
            MOVE -1 TO SOCSECL

        WHEN YRSSVCI EQUAL SPACES OR YRSSVCL EQUAL ZERO
            MOVE "YEARS OF SERVICE IS REQUIRED" TO MESSAGEO
            MOVE -1 TO YRSSVCL

        WHEN YRSSVCI IS NOT NUMERIC
            MOVE "YEARS OF SERVICE MUST BE NUMERIC" TO MESSAGEO
            MOVE -1 TO YRSSVCL

        WHEN LSTPRMI EQUAL SPACES OR LSTPRML EQUAL ZERO
            MOVE "LAST PROMOTION DATE IS REQUIRED" TO MESSAGEO
            MOVE -1 TO LSTPRML

        WHEN OTHER
            PERFORM ADD-RECORD
            MOVE -1 TO EMPNOL

    END-EVALUATE.

    PERFORM SEND-MAP-DATA.
```

```
    ADD-RECORD.

*   MAP INPUT FIELDS TO DB2 RECORD

       MOVE EMPNOI            TO WS-EMPNO
       MOVE WS-EMPNO          TO EMP-ID
       MOVE LNAMEI            TO EMP-LAST-NAME-TEXT

       SET SW-SPACE-NOT-FOUND TO TRUE

       PERFORM VARYING POS-CTR FROM +1 BY +1
          UNTIL (POS-CTR = 30) OR SW-SPACE-FOUND
             IF EMP-LAST-NAME-TEXT(POS-CTR:1) = SPACE
                SET SW-SPACE-FOUND TO TRUE
             END-IF
       END-PERFORM

       IF POS-CTR EQUAL +30
          MOVE +30 TO EMP-LAST-NAME-LEN
       ELSE
          SUBTRACT +1 FROM POS-CTR
          MOVE POS-CTR TO EMP-LAST-NAME-LEN
       END-IF

       MOVE FNAMEI            TO EMP-FIRST-NAME-TEXT

       SET SW-SPACE-NOT-FOUND TO TRUE

       PERFORM VARYING POS-CTR FROM +1 BY +1
          UNTIL (POS-CTR = 20) OR SW-SPACE-FOUND
             IF EMP-FIRST-NAME-TEXT(POS-CTR:1) = SPACE
                SET SW-SPACE-FOUND TO TRUE
             END-IF
       END-PERFORM

       IF POS-CTR EQUAL +20
          MOVE +20 TO EMP-FIRST-NAME-LEN
       ELSE
          SUBTRACT +1 FROM POS-CTR
          MOVE POS-CTR TO EMP-FIRST-NAME-LEN
       END-IF

       MOVE SOCSECI           TO EMP-SSN
       MOVE YRSSVCI           TO WS-EMP-SRV-YRS
       MOVE WS-EMP-SRV-YRS    TO EMP-SERVICE-YEARS
       MOVE LSTPRMI           TO EMP-PROMOTION-DATE

*   INSERT THE RECORD

       EXEC SQL
          INSERT INTO USER01.EMPLOYEE
          (EMP_ID,
           EMP_LAST_NAME,
           EMP_FIRST_NAME,
           EMP_SERVICE_YEARS,
           EMP_PROMOTION_DATE,
           EMP_SSN)
```

```
              VALUES
              (:EMP-ID,
               :EMP-LAST-NAME,
               :EMP-FIRST-NAME,
               :EMP-SERVICE-YEARS,
               :EMP-PROMOTION-DATE,
               :EMP-SSN)

          END-EXEC

 *  HANDLE THE SQLCODE AND RETURN

          MOVE SQLCODE TO WS-SQLCODE

          EVALUATE SQLCODE
              WHEN 0
                  MOVE "EMPLOYEE ADDED SUCCESSFULLY" TO MESSAGEO
                  MOVE -1 TO EMPNOL
                  MOVE WS-EMPNO TO WS-EMP-PASS
              WHEN -803
                  MOVE "ERROR - RECORD ALREADY EXISTS" TO MESSAGEO
                  MOVE -1 TO EMPNOL
                  MOVE WS-EMPNO TO WS-EMP-PASS
              WHEN OTHER
                  STRING "ERROR - DB2 SQLCODE IS " DELIMITED BY SIZE
                     WS-SQLCODE DELIMITED BY SIZE
                         INTO MESSAGEO
                  MOVE -1 TO EMPNOL

          END-EVALUATE.

 BRANCH-TO-PROGRAM.
      EXEC CICS
          XCTL PROGRAM(PROGRAM-NAME)
          COMMAREA (WS-COMMAREA)
          LENGTH(10)
      END-EXEC

      MOVE 'PROGRAM NOT AVAILABLE' TO MESSAGEO.

 SEND-MAP.
      EXEC CICS SEND
          MAP    ('EMPADD')
          MAPSET ('EMPMADD')
          FROM   (EMPADDO)
          CURSOR
          ERASE
      END-EXEC.

 SEND-MAP-DATA.
      EXEC CICS SEND
          MAP    ('EMPADD')
          MAPSET ('EMPMADD')
          FROM   (EMPADDO)
          CURSOR
          DATAONLY
      END-EXEC.
```

```
SEND-MAP-ALL.
    EXEC CICS SEND
        MAP     ('EMPADD')
        MAPSET  ('EMPMADD')
        FROM    (EMPADDO)
        CURSOR
    END-EXEC.

RECEIVE-MAP.
    EXEC CICS RECEIVE
        MAP     ('EMPADD')
        MAPSET  ('EMPMADD')
        INTO    (EMPADDI)
    END-EXEC.
```

Notice in the error handling that if a data value is missing or invalid, we set the length field for that data element to -1. This has the effect of placing the cursor on that screen field. For example:

```
WHEN LNAMEI EQUAL SPACES OR LNAMEL EQUAL ZERO
    MOVE "EMPLOYEE LAST NAME IS REQUIRED" TO MESSAGEO
    MOVE -1 TO LNAMEL
```

Also notice the following code in the data validation routine.

```
MOVE DFHBMFSE TO EMPNOF
MOVE DFHBMFSE TO LNAMEF
MOVE DFHBMFSE TO FNAMEF
MOVE DFHBMFSE TO SOCSECF
MOVE DFHBMFSE TO YRSSVCF
MOVE DFHBMFSE TO LSTPRMF
```

The code turns on the data-changed attribute for each field to ensure that the data is actually sent in the send map routine. Ordinarily you do not need the above code – once you type data into the screen field, the appropriate F variable (such as LNAMEF) will be set by your 3270 emulator program to DFHBMFSE automatically to indicate the field data changed.

I included this code in my program because one of the 3270 emulation products I use has a bug and the data-changed attribute was not getting set. This caused the program to pass a zero length data area which caused an abend. So I forced the data-changed attribute to be set in the program code for each entry field (until I get an updated version of my 3270 emulator). If you ever have this problem, this is one workaround. Hopefully it won't be an issue for you.

Ok, let's get back now and compile, bind and test the program. Make sure you go through all the test cases. My test results are as follows:

```
EMPMADD                    EMPLOYEE ADD                         EMAD


                    ENTER EMPLOYEE INFO, THEN PRESS PF4

          EMPLOYEE ID

          EMP LAST NAME

          EMP FIRST NAME

          EMP SOCIAL SEC

          EMP YEARS SRVC

          EMP LAST PROM




   EMPLOYEE NUMBER IS REQUIRED
   F2 INQ   F3 EXIT   F4 ADD   F5 CHG   F6 DEL



EMPMADD                     EMPLOYEE ADD


                    ENTER EMPLOYEE INFO, THEN PRESS PF4

          EMPLOYEE ID     666N

          EMP LAST NAME

          EMP FIRST NAME

          EMP SOCIAL SEC

          EMP YEARS SRVC

          EMP LAST PROM




   EMPLOYEE NUMBER MUST BE NUMERIC
   F2 INQ   F3 EXIT   F4 ADD   F5 CHG   F6 DEL
```

```
EMPMADD                    EMPLOYEE ADD                         EMAD

               ENTER EMPLOYEE INFO, THEN PRESS PF4

        EMPLOYEE ID    6666

        EMP LAST NAME

        EMP FIRST NAME

        EMP SOCIAL SEC

        EMP YEARS SRVC

        EMP LAST PROM
```

EMPLOYEE LAST NAME IS REQUIRED
```
F2 INQ   F3 EXIT   F4 ADD   F5 CHG   F6 DEL
```

```
EMPMADD                    EMPLOYEE ADD                         EMAD

               ENTER EMPLOYEE INFO, THEN PRESS PF4

        EMPLOYEE ID    6666

        EMP LAST NAME  EDEN

        EMP FIRST NAME

        EMP SOCIAL SEC

        EMP YEARS SRVC

        EMP LAST PROM
```

EMPLOYEE FIRST NAME IS REQUIRED
```
F2 INQ   F3 EXIT   F4 ADD   F5 CHG   F6 DEL
```

```
EMPMADD                         EMPLOYEE ADD                              EMAD


                    ENTER EMPLOYEE INFO, THEN PRESS PF4

          EMPLOYEE ID    6666

          EMP LAST NAME  EDEN

          EMP FIRST NAME HELEN

          EMP SOCIAL SEC

          EMP YEARS SRVC

          EMP LAST PROM




SOCIAL SECURITY NUMBER IS REQUIRED
F2 INQ   F3 EXIT   F4 ADD   F5 CHG   F6 DEL


EMPMADD                         EMPLOYEE ADD                              EMAD


                    ENTER EMPLOYEE INFO, THEN PRESS PF4

          EMPLOYEE ID    6666

          EMP LAST NAME  EDEN

          EMP FIRST NAME HELEN

          EMP SOCIAL SEC 45239866C

          EMP YEARS SRVC

          EMP LAST PROM




SOCIAL SECURITY MUST BE NUMERIC
F2 INQ   F3 EXIT   F4 ADD   F5 CHG   F6 DEL
```

EMPMADD   EMPLOYEE ADD   EMAD

                    ENTER EMPLOYEE INFO, THEN PRESS PF4

        EMPLOYEE ID   6666

        EMP LAST NAME   EDEN

        EMP FIRST NAME HELEN

        EMP SOCIAL SEC 452398667

        EMP YEARS SRVC

        EMP LAST PROM

YEARS OF SERVICE IS REQUIRED
F2 INQ   F3 EXIT   F4 ADD   F5 CHG   F6 DEL

EMPMADD   EMPLOYEE ADD   EMAD

                    ENTER EMPLOYEE INFO, THEN PRESS PF4

        EMPLOYEE ID   6666

        EMP LAST NAME   EDEN

        EMP FIRST NAME HELEN

        EMP SOCIAL SEC 452398667

        EMP YEARS SRVC D7

        EMP LAST PROM

YEARS OF SERVICE MUST BE NUMERIC
F2 INQ   F3 EXIT   F4 ADD   F5 CHG   F6 DEL

```
EMPMADD                    EMPLOYEE ADD                          EMAD


                    ENTER EMPLOYEE INFO, THEN PRESS PF4

         EMPLOYEE ID    6666

         EMP LAST NAME  EDEN

         EMP FIRST NAME HELEN

         EMP SOCIAL SEC 452398667

         EMP YEARS SRVC 17

         EMP LAST PROM
```

LAST PROMOTION DATE IS REQUIRED
```
F2 INQ   F3 EXIT   F4 ADD   F5 CHG   F6 DEL
```


```
EMPMADD                    EMPLOYEE ADD                          EMAD


                    ENTER EMPLOYEE INFO, THEN PRESS PF4

         EMPLOYEE ID    6666

         EMP LAST NAME  EDEN

         EMP FIRST NAME HELEN

         EMP SOCIAL SEC 452398667

         EMP YEARS SRVC 17

         EMP LAST PROM  2018-02-30
```

ERROR - DB2 SQLCODE IS 00000181
```
F2 INQ   F3 EXIT   F4 ADD   F5 CHG   F6 DEL
```

Note: the above error means a bad date. You could have just coded the message, but I wanted to demonstrate an example of formatting an SQL code. For user friendliness, you can change the message to "Last promotion date must be valid" or something like that.

```
EMPMADD                    EMPLOYEE ADD                        EMAD


                  ENTER EMPLOYEE INFO, THEN PRESS PF4

        EMPLOYEE ID    6666

        EMP LAST NAME  EDEN

        EMP FIRST NAME HELEN

        EMP SOCIAL SEC 452398667

        EMP YEARS SRVC 17

        EMP LAST PROM  2018-01-01
```

EMPLOYEE ADDED SUCCESSFULLY
```
F2 INQ   F3 EXIT   F4 ADD   F5 CHG   F6 DEL
```

Ok, we successfully added the record. Now try to add the record again by pressing PF4. You should get an error message that the record already exists.

```
EMPMADD                    EMPLOYEE ADD                        EMAD


                  ENTER EMPLOYEE INFO, THEN PRESS PF4

        EMPLOYEE ID    6666

        EMP LAST NAME  EDEN

        EMP FIRST NAME HELEN

        EMP SOCIAL SEC 452398667

        EMP YEARS SRVC 17

        EMP LAST PROM  2018-01-01
```

ERROR - RECORD ALREADY EXISTS
```
F2 INQ   F3 EXIT   F4 ADD   F5 CHG   F6 DEL
```

Also try an invalid key such as PF7:

```
EMPMADD                         EMPLOYEE ADD                              EMAD


                      ENTER EMPLOYEE INFO, THEN PRESS PF4

          EMPLOYEE ID    1234

          EMP LAST NAME  Test

          EMP FIRST NAME Test

          EMP SOCIAL SEC 999999999

          EMP YEARS SRVC 22

          EMP LAST PROM  2018-01-01




     INVALID KEY PRESSED
     F2 INQ   F3 EXIT   F4 ADD   F5 CHG   F6 DEL
```

Last, press PF3 to end the transaction, then the clear key to clear the screen.

```
EMPMADD                         EMPLOYEE ADD                              EMAD


                      ENTER EMPLOYEE INFO, THEN PRESS PF4

          EMPLOYEE ID    6666

          EMP LAST NAME  EDEN

          EMP FIRST NAME HELEN

          EMP SOCIAL SEC 452398667

          EMP YEARS SRVC 17

          EMP LAST PROM  2018-01-01




     BYE, PRESS CLEAR KEY TO ENTER A TRANSACTION ID
     F2 INQ   F3 EXIT   F4 ADD   F5 CHG   F6 DEL
```

Ok that's it for the add program. Let's move on to the change program, EMPPGCHG.

Employee Change Program

The following are specifications for our change program. Note that there are two SQL statements, one to retrieve the record and one to update it.

Name: EMPPGCHG

Purpose: Change an employee in the EMPLOYEE table.

Data Access: Read and Update on EMPLOYEE

SQL Statements:

```
        SELECT EMP_ID,
               EMP_LAST_NAME,
               EMP_FIRST_NAME,
               EMP_SSN,
               EMP_SERVICE_YEARS,
               EMP_PROMOTION_DATE
          INTO
               :EMP-ID,
               :EMP-LAST-NAME,
               :EMP-FIRST-NAME,
               :EMP-SSN,
               :EMP-SERVICE-YEARS,
               :EMP-PROMOTION-DATE
          FROM USER01.EMPLOYEE
          WHERE EMP_ID = :EMP-ID


UPDATE USER01.EMPLOYEE
SET EMP_LAST_NAME       = :EMP-LAST-NAME,
    EMP_FIRST_NAME      = :EMP-FIRST-NAME,
    EMP_SERVICE_YEARS   = :EMP-SERVICE-YEARS,
    EMP_PROMOTION_DATE  = :EMP-PROMOTION-DATE,
    EMP_SSN             = :EMP-SSN
WHERE EMP_ID = :EMP-ID
```

EMPPGCHG Test Cases:

Case	Condition	Expected Result	Actual Result
1	Initial screen display	Literals displayed, all enterable fields unprotected.	
2	Valid employee number	Display detail for employee	
3	Invalid employee number	Error message that emp number is invalid	
4	Enter pressed without changing anything	No change	
5	Employee number is blank	Error – employee number is required	
6	Employee number not numeric	Error – employee number must be numeric	
7	Last name is blank	Error – Last Name is required	

8	First name is blank	Error – Fast Name is required	
9	Social Security number is blank	Error – Social Security Number is required	
10	Social Security number is not numeric	Error – Social Security Number must be numeric	
11	Years of Service is blank	Error – Years of Service is required	
12	Years of Service not numeric	Error – Years of Service must be numeric	
13	Last Promotion Date is blank	Error – Last Promotion Date is required	
14	PF2 pressed	Transfer to EMIN transaction	
15	PF3 pressed	Message to clear screen and enter a transaction id	
16	PF4 pressed	Transfer to EMAD transaction	
17	PF5 pressed	Change the record if no errors, message that record was changed	
18	PF6 Pressed	Transfer to EMDE transaction	
19	Other PF Keys pressed (PF1, PF7, PF8)	Error message – invalid key	

Employee Change Program Code

Here is the code for the change program. Much of the program looks like the add program, except we also do an initial display of the record before we edit and update it. Please review this change program carefully to see how we handle field edits and errors. Also notice that the "action" key for this transaction is PF5 (not ENTER, although you use ENTER if you are keying in a new employee id for display).

```
IDENTIFICATION DIVISION.
PROGRAM-ID. EMPPGCHG.

***************************************************
*   COBOL/CICS/DB2 PROGRAM TO CHANGE AN EMPLOYEE  *
*                                                 *
*   AUTHOR       : ROBERT WINGATE                 *
*   DATE-WRITTEN : 2018-07-23                     *
***************************************************

ENVIRONMENT DIVISION.

DATA DIVISION.

WORKING-STORAGE SECTION.
01 WS-EMPNO        PIC 9(4).
01 WS-EMP-SRV-YRS  PIC 9(02).
01 WS-SQLCODE      PIC 9(08).
01 WS-COMMAREA.
   05 WS-EMP-PASS  PIC 9(04).
   05 WS-PGM-PASS  PIC X(08).
   05 FILLER       PIC X(08).

01 PROGRAM-NAME    PIC X(08) VALUE SPACES.
```

```cobol
01 POS-CTR             PIC S9(9) USAGE COMP VALUE +0.

01 SW-SPACE-FOUND-SWITCH   PIC X(1) VALUE 'N'.
   88  SW-SPACE-FOUND                 VALUE 'Y'.
   88  SW-SPACE-NOT-FOUND             VALUE 'N'.

01 SW-PASSED-DATA-SWITCH   PIC X(1) VALUE 'N'.
   88  SW-PASSED-DATA                 VALUE 'Y'.
   88  SW-NO-PASSED-DATA              VALUE 'N'.

   COPY EMPMCHG.
   COPY DFHAID.
   COPY DFHBMSCA.

   EXEC SQL
     INCLUDE SQLCA
   END-EXEC.

   EXEC SQL
     INCLUDE EMPLOYEE
   END-EXEC.

LINKAGE SECTION.

01 DFHCOMMAREA          PIC X(20).

PROCEDURE DIVISION.

   SET SW-NO-PASSED-DATA TO TRUE

   IF EIBCALEN > ZERO
     MOVE DFHCOMMAREA   TO WS-COMMAREA
   END-IF.

   EVALUATE TRUE

     WHEN EIBCALEN = ZERO
       MOVE LOW-VALUES    TO   EMPCHGO
       MOVE -1 TO EMPINL
       PERFORM SEND-MAP

     WHEN EIBAID = DFHCLEAR
       MOVE LOW-VALUES    TO   EMPCHGO
       MOVE -1 TO EMPINL
       PERFORM SEND-MAP

     WHEN EIBAID = DFHPA1 OR DFHPA2 OR DFHPA3
       CONTINUE

     WHEN EIBAID = DFHPF2
       MOVE 'EMPPGINQ' TO PROGRAM-NAME
       PERFORM BRANCH-TO-PROGRAM

       EXEC CICS
         RETURN
       END-EXEC
```

```
            WHEN EIBAID = DFHPF3
              MOVE LOW-VALUES TO  EMPCHGO
              MOVE -1 TO EMPINL
              MOVE "BYE, PRESS CLEAR KEY TO ENTER A TRANSACTION ID"
                   TO MESSAGEO
              PERFORM SEND-MAP-DATA

              EXEC CICS
                RETURN
              END-EXEC

            WHEN EIBAID = DFHPF4
              MOVE 'EMPPGADD' TO PROGRAM-NAME
              PERFORM BRANCH-TO-PROGRAM

              EXEC CICS
                RETURN
              END-EXEC

            WHEN EIBAID = DFHPF5
*             PERFORM THE EDITS AND VALIDATIONS
*             IF NO ERRORS THEN MODIFY THE RECORD

              IF WS-PGM-PASS NOT EQUAL "EMPPGCHG"
                 SET SW-PASSED-DATA TO TRUE
                 PERFORM PROCESS-PARA
              ELSE
                 PERFORM VALIDATE-DATA
              END-IF

            WHEN EIBAID = DFHPF6
              MOVE 'EMPPGDEL' TO PROGRAM-NAME
              PERFORM BRANCH-TO-PROGRAM

              EXEC CICS
                RETURN
              END-EXEC

            WHEN EIBAID = DFHENTER
              PERFORM PROCESS-PARA

            WHEN OTHER
              MOVE LOW-VALUES TO EMPCHGO
              MOVE -1 TO EMPINL
              MOVE "INVALID KEY PRESSED" TO MESSAGEO
              PERFORM SEND-MAP-DATA

        END-EVALUATE.

        EXEC CICS
           RETURN TRANSID('EMCH')
           COMMAREA (WS-COMMAREA)
           LENGTH(20)
        END-EXEC.

    PROCESS-PARA.
```

```
PERFORM RECEIVE-MAP.
INITIALIZE DCLEMPLOYEE MESSAGEO

IF SW-PASSED-DATA
    MOVE WS-EMP-PASS TO EMP-ID
ELSE
    MOVE EMPINI     TO WS-EMPNO WS-EMP-PASS
    MOVE WS-EMPNO   TO EMP-ID
END-IF

MOVE DFHBMFSE TO EMPINF
MOVE DFHBMFSE TO EMPNOF
MOVE DFHBMFSE TO LNAMEF
MOVE DFHBMFSE TO FNAMEF
MOVE DFHBMFSE TO SOCSECF
MOVE DFHBMFSE TO YRSSVCF
MOVE DFHBMFSE TO LSTPRMF

EXEC SQL
    SELECT EMP_ID,
           EMP_LAST_NAME,
           EMP_FIRST_NAME,
           EMP_SSN,
           EMP_SERVICE_YEARS,
           EMP_PROMOTION_DATE
      INTO
           :EMP-ID,
           :EMP-LAST-NAME,
           :EMP-FIRST-NAME,
           :EMP-SSN,
           :EMP-SERVICE-YEARS,
           :EMP-PROMOTION-DATE
      FROM USER01.EMPLOYEE
      WHERE EMP_ID = :EMP-ID
END-EXEC.

MOVE SQLCODE        TO  WS-SQLCODE.
DISPLAY "SQLCODE: " WS-SQLCODE.

EVALUATE SQLCODE
  WHEN 0
     MOVE EMP-ID               TO WS-EMPNO WS-EMP-PASS
     MOVE WS-EMPNO             TO EMPNOO EMPINO
     MOVE EMP-LAST-NAME-TEXT   TO LNAMEO
     MOVE EMP-FIRST-NAME-TEXT  TO FNAMEO
     MOVE EMP-SSN              TO SOCSECO
     MOVE EMP-SERVICE-YEARS    TO WS-EMP-SRV-YRS
     MOVE WS-EMP-SRV-YRS       TO YRSSVCO
     MOVE EMP-PROMOTION-DATE   TO LSTPRMO

     MOVE "MAKE CHANGES AND THEN PRESS PF5" TO MESSAGEO

  WHEN 100
     STRING "EMPLOYEE ID " DELIMITED BY SIZE
     WS-EMPNO DELIMITED BY SPACE
     " NOT FOUND" DELIMITED BY SIZE INTO MESSAGEO
     MOVE WS-EMPNO      TO EMPINO
     MOVE SPACES        TO EMPNOO
```

```
                MOVE SPACES         TO LNAMEO
                MOVE SPACES         TO FNAMEO
                MOVE SPACES         TO SOCSECO
                MOVE SPACES         TO YRSSVCO
                MOVE SPACES         TO LSTPRMO

            WHEN OTHER
                STRING "SQL CODE: " DELIMITED BY SIZE
                       WS-SQLCODE   DELIMITED BY SIZE
                   INTO MESSAGEO

        END-EVALUATE.

        MOVE DFHBMFSE   TO EMPINF
        MOVE -1 TO EMPINL.
        MOVE "EMPPGCHG" TO WS-PGM-PASS

        IF SW-PASSED-DATA
           PERFORM SEND-MAP
        ELSE
           PERFORM SEND-MAP-DATA
        END-IF.

    VALIDATE-DATA.

        PERFORM RECEIVE-MAP
        INITIALIZE DCLEMPLOYEE MESSAGEO

        EVALUATE TRUE

            WHEN EMPNOI EQUAL SPACES OR EMPNOL EQUAL ZERO
                MOVE "EMPLOYEE NUMBER IS REQUIRED" TO MESSAGEO
                MOVE -1 TO EMPNOL

            WHEN EMPNOI IS NOT NUMERIC
                MOVE "EMPLOYEE NUMBER MUST BE NUMERIC" TO MESSAGEO
                MOVE -1 TO EMPNOL

            WHEN LNAMEI EQUAL SPACES OR LNAMEL EQUAL ZERO
                MOVE "EMPLOYEE LAST NAME IS REQUIRED" TO MESSAGEO
                MOVE -1 TO LNAMEL

            WHEN FNAMEI EQUAL SPACES OR FNAMEL EQUAL ZERO
                MOVE "EMPLOYEE FIRST NAME IS REQUIRED" TO MESSAGEO
                MOVE -1 TO FNAMEL

            WHEN SOCSECI EQUAL SPACES OR SOCSECL EQUAL ZERO
                MOVE "SOCIAL SECURITY NUMBER IS REQUIRED" TO MESSAGEO
                MOVE -1 TO SOCSECL

            WHEN SOCSECI IS NOT NUMERIC
                MOVE "SOCIAL SECURITY MUST BE NUMERIC" TO MESSAGEO
                MOVE -1 TO SOCSECL

            WHEN YRSSVCI EQUAL SPACES OR YRSSVCL EQUAL ZERO
                MOVE "YEARS OF SERVICE IS REQUIRED" TO MESSAGEO
                MOVE -1 TO YRSSVCL
```

```
        WHEN YRSSVCI IS NOT NUMERIC
            MOVE "YEARS OF SERVICE MUST BE NUMERIC" TO MESSAGEO
            MOVE -1 TO YRSSVCL

        WHEN LSTPRMI EQUAL SPACES OR LSTPRML EQUAL ZERO
            MOVE "LAST PROMOTION DATE IS REQUIRED" TO MESSAGEO
            MOVE -1 TO LSTPRML

        WHEN OTHER
            PERFORM CHANGE-RECORD
            MOVE -1 TO EMPINL

    END-EVALUATE.

    MOVE DFHBMFSE TO EMPINF
    MOVE DFHBMFSE TO EMPNOF
    MOVE DFHBMFSE TO LNAMEF
    MOVE DFHBMFSE TO FNAMEF
    MOVE DFHBMFSE TO SOCSECF
    MOVE DFHBMFSE TO YRSSVCF
    MOVE DFHBMFSE TO LSTPRMF

    MOVE "EMPPGCHG" TO WS-PGM-PASS
    PERFORM SEND-MAP-DATA.

CHANGE-RECORD.

*   MAP INPUT FIELDS TO DB2 RECORD

    MOVE EMPNOI              TO WS-EMPNO
    MOVE WS-EMPNO            TO EMP-ID
    MOVE LNAMEI              TO EMP-LAST-NAME-TEXT

    SET SW-SPACE-NOT-FOUND TO TRUE

    PERFORM VARYING POS-CTR FROM +1 BY +1
        UNTIL (POS-CTR = 30) OR SW-SPACE-FOUND
            IF EMP-LAST-NAME-TEXT(POS-CTR:1) = SPACE
                SET SW-SPACE-FOUND TO TRUE
            END-IF
    END-PERFORM

    IF POS-CTR EQUAL +30
        MOVE +30 TO EMP-LAST-NAME-LEN
    ELSE
        SUBTRACT +1 FROM POS-CTR
        MOVE POS-CTR TO EMP-LAST-NAME-LEN
    END-IF

    MOVE FNAMEI             TO EMP-FIRST-NAME-TEXT

    SET SW-SPACE-NOT-FOUND TO TRUE

    PERFORM VARYING POS-CTR FROM +1 BY +1
        UNTIL (POS-CTR = 20) OR SW-SPACE-FOUND
            IF EMP-FIRST-NAME-TEXT(POS-CTR:1) = SPACE
                SET SW-SPACE-FOUND TO TRUE
            END-IF
```

```
            END-PERFORM

        IF POS-CTR EQUAL +20
           MOVE +20 TO EMP-FIRST-NAME-LEN
        ELSE
           SUBTRACT +1 FROM POS-CTR
           MOVE POS-CTR TO EMP-FIRST-NAME-LEN
        END-IF

        MOVE SOCSECI           TO EMP-SSN
        MOVE YRSSVCI           TO WS-EMP-SRV-YRS
        MOVE WS-EMP-SRV-YRS    TO EMP-SERVICE-YEARS
        MOVE LSTPRMI           TO EMP-PROMOTION-DATE

*   UPDATE THE RECORD

        EXEC SQL
           UPDATE USER01.EMPLOYEE
           SET EMP_LAST_NAME      = :EMP-LAST-NAME,
               EMP_FIRST_NAME     = :EMP-FIRST-NAME,
               EMP_SERVICE_YEARS  = :EMP-SERVICE-YEARS,
               EMP_PROMOTION_DATE = :EMP-PROMOTION-DATE,
               EMP_SSN            = :EMP-SSN
           WHERE EMP_ID = :EMP-ID

        END-EXEC

* HANDLE THE SQLCODE AND RETURN

        MOVE SQLCODE TO WS-SQLCODE

        EVALUATE SQLCODE
           WHEN 0
              MOVE "EMPLOYEE MODIFIED SUCCESSFULLY" TO MESSAGEO
              MOVE -1 TO EMPNOL
           WHEN +100
              MOVE "ERROR - RECORD NOT FOUND" TO MESSAGEO
              MOVE -1 TO EMPNOL
           WHEN OTHER
              STRING "ERROR - DB2 SQLCODE IS " DELIMITED BY SIZE
                 WS-SQLCODE DELIMITED BY SIZE
                    INTO MESSAGEO
              MOVE -1 TO EMPNOL

        END-EVALUATE.

  BRANCH-TO-PROGRAM.

        EXEC CICS
           XCTL PROGRAM(PROGRAM-NAME)
           COMMAREA (WS-COMMAREA)
           LENGTH(20)
        END-EXEC

        MOVE 'PROGRAM NOT AVAILABLE' TO MESSAGEO.

  SEND-MAP.
        EXEC CICS SEND
```

```
          MAP      ('EMPCHG')
          MAPSET ('EMPMCHG')
          FROM     (EMPCHGO)
          CURSOR
          ERASE
      END-EXEC.

 SEND-MAP-DATA.
      EXEC CICS SEND
          MAP      ('EMPCHG')
          MAPSET ('EMPMCHG')
          FROM     (EMPCHGO)
          DATAONLY
          CURSOR
      END-EXEC.

 SEND-MAP-ALL.
      EXEC CICS SEND
          MAP      ('EMPCHG')
          MAPSET ('EMPMCHG')
          FROM     (EMPCHGO)
          CURSOR
      END-EXEC.

 RECEIVE-MAP.
      EXEC CICS RECEIVE
          MAP      ('EMPCHG')
          MAPSET ('EMPMCHG')
          INTO     (EMPCHGI)
      END-EXEC.
```

The edits and validations are the same as for the add program. Also for error handling if a data value is missing or invalid, we set the length field for that data element to -1. This has the effect of placing the cursor on that screen field. For example:

```
WHEN LNAMEI EQUAL SPACES OR LNAMEL EQUAL ZERO
    MOVE "EMPLOYEE LAST NAME IS REQUIRED" TO MESSAGEO
    MOVE -1 TO LNAMEL
```

Also notice that we check to see whether another program has passed data to this one, and if it has we use that passed employee id to populate the initial screen display with data. Otherwise we simply display an empty screen.

Now you can go ahead and compile, link and bind the program. Next, define and install the program in CICS. Remember to include a DB2ENTRY resource definition or your program won't work. Finally, run the test cases to make sure all functions work as intended.

When you've finished testing the Change program, let's move on to the Delete program.

Employee Delete Program

The following are specifications for our delete program. Again there are two SQL statements, one to retrieve the record for display, and the other to delete it.

Name: EMPPGDEL
Purpose: Delete an employee from the EMPLOYEE table.
Data Access: Read and Delete on EMPLOYEE
SQL Statements:

```
SELECT EMP_ID,
       EMP_LAST_NAME,
       EMP_FIRST_NAME,
       EMP_SSN,
       EMP_SERVICE_YEARS,
       EMP_PROMOTION_DATE
  INTO
       :EMP-ID,
       :EMP-LAST-NAME,
       :EMP-FIRST-NAME,
       :EMP-SSN,
       :EMP-SERVICE-YEARS,
       :EMP-PROMOTION-DATE
  FROM USER01.EMPLOYEE
  WHERE EMP_ID = :EMP-ID

DELETE FROM USER01.EMPLOYEE
   WHERE EMP_ID = :EMP-ID
```

EMPPGDEL Test Cases:

Case	Condition	Expected Result	Actual Result
1	Initial screen display	Literals displayed, all enterable fields unprotected.	
2	Valid employee number entered	Display detail for employee	
3	Invalid employee number entered	Error message that emp number is invalid	
4	Enter pressed without changing anything	No change	
5	PF2 pressed	Transfer to EMIN transaction	
6	PF3 pressed	Message to clear screen and enter a transaction id	
7	PF4 pressed	Transfer to EMAD transaction	
8	PF5 pressed	Transfer to EMCH transaction	
9	PF6 Pressed	Delete the record if no errors, message that record was deleted	
10	Other PF Keys pressed	Error message – invalid key	

Employee Delete Program Code

Here is the code for the delete program. This one is very much like the change program where we do an initial display of the record before we delete it. Please review this program carefully. Also notice that the "action" key for this transaction is PF6 (not ENTER).

```
        IDENTIFICATION DIVISION.
        PROGRAM-ID. EMPPGDEL.

      ****************************************************
      *   COBOL/CICS/DB2 PROGRAM TO DELETE AN EMPLOYEE  *
      *                                                 *
      *   AUTHOR        : ROBERT WINGATE                *
      *   DATE-WRITTEN  : 2018-07-25                    *
      ****************************************************

        ENVIRONMENT DIVISION.

        DATA DIVISION.

        WORKING-STORAGE SECTION.
        01 WS-EMPNO         PIC 9(4).
        01 WS-EMP-SRV-YRS   PIC 9(2).
        01 WS-SQLCODE       PIC -9(08).
        01 WS-COMMAREA.
           05 WS-EMP-PASS   PIC 9(04).
           05 WS-PGM-PASS   PIC X(08).
           05 FILLER        PIC X(08).


        01 PROGRAM-NAME     PIC X(08) VALUE SPACES.

        01 SW-PASSED-DATA-SWITCH   PIC X(1) VALUE 'N'.
           88  SW-PASSED-DATA               VALUE 'Y'.
           88  SW-NO-PASSED-DATA            VALUE 'N'.

           COPY EMPMDEL.
           COPY DFHAID.
           COPY DFHBMSCA.

           EXEC SQL
             INCLUDE SQLCA
           END-EXEC.

           EXEC SQL
             INCLUDE EMPLOYEE
           END-EXEC.

        LINKAGE SECTION.

        01 DFHCOMMAREA         PIC X(20).

        PROCEDURE DIVISION.

           IF EIBCALEN > ZERO
             MOVE DFHCOMMAREA  TO WS-COMMAREA
           END-IF.
```

```
EVALUATE TRUE

   WHEN EIBCALEN = ZERO
      MOVE LOW-VALUES   TO   EMPDELO
      PERFORM SEND-MAP

   WHEN EIBAID = DFHCLEAR
      MOVE LOW-VALUES   TO   EMPDELO
      PERFORM SEND-MAP

   WHEN EIBAID = DFHPA1 OR DFHPA2 OR DFHPA3
      CONTINUE

   WHEN EIBAID = DFHPF2
      MOVE 'EMPPGINQ' TO PROGRAM-NAME
      PERFORM BRANCH-TO-PROGRAM

   WHEN EIBAID = DFHPF3
      MOVE LOW-VALUES TO  EMPDELO
      MOVE "BYE, PRESS CLEAR KEY TO ENTER A TRANSACTION ID"
          TO MESSAGEO
      PERFORM SEND-MAP-DATA

      EXEC CICS
         RETURN
      END-EXEC

   WHEN EIBAID = DFHPF4
      MOVE 'EMPPGADD' TO PROGRAM-NAME
      PERFORM BRANCH-TO-PROGRAM

   WHEN EIBAID = DFHPF5
      MOVE 'EMPPGCHG' TO PROGRAM-NAME
      PERFORM BRANCH-TO-PROGRAM

   WHEN EIBAID = DFHPF6

      IF WS-PGM-PASS NOT EQUAL "EMPPGDEL"
         SET SW-PASSED-DATA TO TRUE
         PERFORM PROCESS-PARA
      ELSE
         SET SW-NO-PASSED-DATA TO TRUE
         PERFORM DELETE-RECORD
      END-IF

   WHEN EIBAID = DFHENTER
      PERFORM PROCESS-PARA

   WHEN OTHER
      MOVE LOW-VALUES TO EMPDELO
      MOVE "INVALID KEY PRESSED" TO MESSAGEO
      PERFORM SEND-MAP-DATA

END-EVALUATE.

EXEC CICS
   RETURN TRANSID('EMDE')
```

```
        COMMAREA (WS-COMMAREA)
    END-EXEC.

PROCESS-PARA.

    PERFORM RECEIVE-MAP.
    INITIALIZE DCLEMPLOYEE MESSAGEO

    IF SW-PASSED-DATA
       MOVE WS-EMP-PASS TO EMP-ID
    ELSE
       MOVE EMPINI    TO WS-EMPNO WS-EMP-PASS
       MOVE WS-EMPNO  TO EMP-ID
    END-IF

    EXEC SQL
       SELECT EMP_ID,
              EMP_LAST_NAME,
              EMP_FIRST_NAME,
              EMP_SSN,
              EMP_SERVICE_YEARS,
              EMP_PROMOTION_DATE
          INTO
              :EMP-ID,
              :EMP-LAST-NAME,
              :EMP-FIRST-NAME,
              :EMP-SSN,
              :EMP-SERVICE-YEARS,
              :EMP-PROMOTION-DATE
          FROM USER01.EMPLOYEE
          WHERE EMP_ID = :EMP-ID
    END-EXEC.

    MOVE SQLCODE        TO  WS-SQLCODE.

    EVALUATE SQLCODE
      WHEN 0
        MOVE EMP-ID                TO  WS-EMPNO
        MOVE WS-EMPNO              TO  EMPNOO EMPINO
        MOVE EMP-LAST-NAME-TEXT   TO LNAMEO
        MOVE EMP-FIRST-NAME-TEXT  TO FNAMEO
        MOVE EMP-SSN               TO SOCSECO
        MOVE EMP-SERVICE-YEARS    TO WS-EMP-SRV-YRS
        MOVE WS-EMP-SRV-YRS       TO YRSSVCO
        MOVE EMP-PROMOTION-DATE   TO LSTPRMO
        MOVE "EMPPGDEL"            TO WS-PGM-PASS

        MOVE "PRESS PF6 TO DELETE EMPLOYEE" TO MESSAGEO

      WHEN 100
        STRING "EMPLOYEE ID " DELIMITED BY SIZE
        WS-EMPNO DELIMITED BY SPACE
        " NOT FOUND" DELIMITED BY SIZE INTO MESSAGEO
        MOVE WS-EMPNO      TO EMPINO
        MOVE SPACES        TO EMPNOO
        MOVE SPACES        TO LNAMEO
        MOVE SPACES        TO FNAMEO
        MOVE SPACES        TO SOCSECO
```

111

```
                MOVE SPACES          TO YRSSVCO
                MOVE SPACES          TO LSTPRMO

           WHEN OTHER
             STRING "SQL CODE: " DELIMITED BY SIZE
                    WS-SQLCODE   DELIMITED BY SIZE
                INTO MESSAGEO
        END-EVALUATE.

        MOVE DFHBMFSE TO EMPINF
        MOVE -1 TO EMPINL.
        MOVE "EMPPGDEL" TO WS-PGM-PASS

        IF SW-PASSED-DATA
           PERFORM SEND-MAP
        ELSE
           PERFORM SEND-MAP-DATA
        END-IF.

    DELETE-RECORD.

        PERFORM RECEIVE-MAP.
        INITIALIZE DCLEMPLOYEE MESSAGEO

*   MAP INPUT FIELDS TO DB2 RECORD

        MOVE EMPINI          TO WS-EMPNO
        MOVE WS-EMPNO         TO EMP-ID

*   DELETE THE RECORD

        EXEC SQL
           DELETE FROM USER01.EMPLOYEE
              WHERE EMP_ID = :EMP-ID
        END-EXEC

*   HANDLE THE SQLCODE AND RETURN

        MOVE SQLCODE TO WS-SQLCODE

        EVALUATE SQLCODE
           WHEN 0
              MOVE "EMPLOYEE DELETED SUCCESSFULLY" TO MESSAGEO
              MOVE -1 TO EMPNOL
              MOVE WS-EMPNO TO WS-EMP-PASS
           WHEN +100
              MOVE "ERROR - RECORD DOES NOT EXIST" TO MESSAGEO
              MOVE -1 TO EMPNOL
              MOVE WS-EMPNO TO WS-EMP-PASS
           WHEN OTHER
              STRING "ERROR - DB2 SQLCODE IS " DELIMITED BY SPACE
                 WS-SQLCODE DELIMITED BY SPACE
                    INTO MESSAGEO
              MOVE -1 TO EMPNOL

        END-EVALUATE.

        MOVE DFHBMFSE TO EMPINF
```

112

```
              MOVE -1 TO EMPINL.
              MOVE "EMPPGDEL" TO WS-PGM-PASS
              PERFORM SEND-MAP-DATA.

       BRANCH-TO-PROGRAM.

              EXEC CICS
                  XCTL PROGRAM(PROGRAM-NAME)
                  COMMAREA (WS-COMMAREA)
                  LENGTH(10)
              END-EXEC

              MOVE 'PROGRAM NOT AVAILABLE' TO MESSAGEO.

       SEND-MAP.
              EXEC CICS SEND
                  MAP    ('EMPDEL')
                  MAPSET ('EMPMDEL')
                  FROM   (EMPDELO)
                  ERASE
              END-EXEC.

       SEND-MAP-DATA.
              EXEC CICS SEND
                  MAP    ('EMPDEL')
                  MAPSET ('EMPMDEL')
                  FROM   (EMPDELO)
                  DATAONLY
              END-EXEC.

       SEND-MAP-ALL.
              EXEC CICS SEND
                  MAP    ('EMPDEL')
                  MAPSET ('EMPMDEL')
                  FROM   (EMPDELO)
              END-EXEC.

       RECEIVE-MAP.
              EXEC CICS RECEIVE
                  MAP    ('EMPDEL')
                  MAPSET ('EMPMDEL')
                  INTO   (EMPDELI)
              END-EXEC.
```

Now you can go ahead and compile, link and bind the program. Next, define and install the program in CICS. Remember to include a DB2ENTRY resource definition or your program won't work. Finally, run the test cases to make sure all functions work as intended.

When you've finished testing the Delete program, let's circle back to the Menu program.

Employee Support Menu

Now we're back to our menu program which should work now because the other programs exist and we can transfer to them. Go ahead and compile, link, bind and test the menu program.

Program: EMPPGMNU
Purpose: Driver for employee support functions
Data Access: None

EMPPGMNU Test Cases

Here are the test cases for the menu program. Since the four function programs have been created we can now test the menu program.

Case	Condition	Expected Result	Actual Result
1	Initial screen display	All literals displayed, empty action field	
2	Invalid selection entered	Error message returned: ENTER A VALID AC-TION: 1, 2, 3 OR 4	
3	Invalid key pressed	Error message: INVALID KEY PRESSED	
4	PF3 pressed	Message to clear screen and enter a trans id	
5	Option 1 entered	Transfer to EMIN transaction	
6	Option 2 entered	Transfer to EMAD transaction	
7	Option 3 entered	Transfer to EMCH transaction	
8	Option 4 entered	Transfer to EMDE transaction	

Employee Support Menu Program Code

We've already provided the code for the menu program earlier, but to keep everything together, we'll repeat it here.

```
        IDENTIFICATION DIVISION.
        PROGRAM-ID. EMPPGMNU.

        ***************************************************
        *   COBOL/CICS EMPLOYEE SUPPORT MENU PROGRAM      *
        *                                                 *
        *   AUTHOR        : ROBERT WINGATE                *
        *   DATE-WRITTEN  : 2018-07-26                    *
        ***************************************************

        ENVIRONMENT DIVISION.
```

```
DATA DIVISION.

WORKING-STORAGE SECTION.

01 WS-FLAGS.
    05 SW-VALID-SELECTION       PIC X(1) VALUE 'N'.
        88  VALID-SELECTION              VALUE 'Y'.
        88  NOT-VALID-SELECTION          VALUE 'N'.

    05 SW-SEND-FLAGS           PIC X(1).
        88  SEND-ERASE                   VALUE '1'.
        88  SEND-DATAONLY                VALUE '2'.
        88  SEND-DATAONLY-ALARM          VALUE '3'.

01 WS-VARS.
    05 COMM-AREA               PIC X(20) VALUE SPACE.
    05 WS-SQLCODE              PIC -9(08).
    05 PROGRAM-NAME            PIC X(08) VALUE SPACES.
    05 RESPONSE-CODE           PIC S9(08) COMP.
    05 INVALID-ACTION-MSG      PIC X(31)
       VALUE 'ENTER A VALID ACTION: 1, 2 OR 3'.
    05 END-SESSION-MSG         PIC X(23)
       VALUE 'THIS SESSION HAS ENDED.'.

    COPY EMPMMNU.
    COPY DFHAID.
    COPY DFHBMSCA.

LINKAGE SECTION.

01 DFHCOMMAREA          PIC X(10).

PROCEDURE DIVISION.

    IF EIBCALEN > ZERO
      MOVE DFHCOMMAREA  TO COMM-AREA
    END-IF.

    EVALUATE TRUE

      WHEN EIBCALEN = ZERO
        MOVE LOW-VALUES   TO  EMPMNUO
        SET SEND-ERASE TO TRUE
        PERFORM SEND-MAP

      WHEN EIBAID = DFHCLEAR
        MOVE LOW-VALUES    TO  EMPMNUO
        SET SEND-ERASE TO TRUE
        PERFORM SEND-MAP

      WHEN EIBAID = DFHPA1 OR DFHPA2 OR DFHPA3
        CONTINUE

      WHEN EIBAID = DFHPF3
        MOVE LOW-VALUES TO  EMPMNUO
        MOVE "BYE, PRESS CLEAR KEY TO ENTER A TRANSACTION ID"
             TO MESSAGEO
        PERFORM SEND-MAP-DATAONLY
```

```
            EXEC CICS
               RETURN
            END-EXEC

        WHEN EIBAID = DFHENTER
           PERFORM MAIN-MAIN-PROCESS-PARA

        WHEN OTHER
           MOVE LOW-VALUES TO EMPMNUO
           MOVE "INVALID KEY PRESSED" TO MESSAGEO
           PERFORM SEND-MAP-DATAONLY

     END-EVALUATE.

     EXEC CICS
        RETURN TRANSID('EMNU')
        COMMAREA (COMM-AREA)
     END-EXEC.


 MAIN-MAIN-PROCESS-PARA.

     PERFORM RECEIVE-MAP.

     IF ACTIONI NOT = '1' AND '2' AND '3' AND '4'
        MOVE DFHREVRS TO ACTIONH
        MOVE INVALID-ACTION-MSG TO MESSAGEO
        SET NOT-VALID-SELECTION TO TRUE
     ELSE
        SET VALID-SELECTION TO TRUE
     END-IF.

     IF VALID-SELECTION
        EVALUATE ACTIONI
           WHEN '1'
              MOVE 'EMPPGINQ' TO PROGRAM-NAME
           WHEN '2'
              MOVE 'EMPPGADD' TO PROGRAM-NAME
           WHEN '3'
              MOVE 'EMPPGCHG' TO PROGRAM-NAME
           WHEN '4'
              MOVE 'EMPPGDEL' TO PROGRAM-NAME
        END-EVALUATE

        PERFORM BRANCH-TO-PROGRAM

     END-IF.

     SET SEND-DATAONLY-ALARM TO TRUE.
     PERFORM SEND-MAP-DATAONLY-ALARM.

 BRANCH-TO-PROGRAM.

     EXEC CICS
        XCTL PROGRAM(PROGRAM-NAME)
     END-EXEC
```

116

```
            MOVE 'PROGRAM NOT AVAILABLE' TO MESSAGEO.

    SEND-MAP.
        EXEC CICS SEND
            MAP     ('EMPMNU')
            MAPSET ('EMPMMNU')
            FROM    (EMPMNUO)
            ERASE
        END-EXEC.

    SEND-MAP-DATAONLY.
        EXEC CICS SEND
            MAP     ('EMPMNU')
            MAPSET ('EMPMMNU')
            FROM    (EMPMNUO)
            DATAONLY
        END-EXEC.

    SEND-MAP-DATAONLY-ALARM.
        EXEC CICS SEND
            MAP     ('EMPMNU')
            MAPSET ('EMPMMNU')
            FROM    (EMPMNUO)
            DATAONLY
            ALARM
        END-EXEC.

    RECEIVE-MAP.
        EXEC CICS RECEIVE
            MAP     ('EMPMNU')
            MAPSET ('EMPMMNU')
            INTO    (EMPMNUI)
        END-EXEC.
```

Once you have a good unit test, let's proceed to integration testing.

Integration Testing

Now we must test to make sure transfers from one program to the other work correctly, and that the information that needs to be passed in the communication area does in fact get passed. So execute all the program transfers, and then also retest the individual programs to make sure they work correctly when simply entering data on the screen (without data being passed from another program).

My integration test results follow.

EMPPGMU

Ok, begin with the menu program `EMPPGMNU`. We'll simply verify that we can navigate to all the screens and that they set up properly. We'll choose the menu options for inquiry, add, change and delete. First inquiry.

```
EMPMMNU                    EMPLOYEE SUPPORT MENU                    EMNU

          ENTER THE NUMBER OF YOUR SELECTION,  THEN PRESS ENTER.


                    1   1. EMPLOYEE INQUIRY

                        2. EMPLOYEE ADD

                        3. EMPLOYEE CHANGE

                        4. EMPLOYEE DELETE








F3 EXIT


EMPMINQ                    EMPLOYEE INQUIRY                        EMIN

        EMPLOYEE ->          ENTER EMPLOYEE ID, THEN PRESS ENTER

        EMPLOYEE ID   XXXX

        EMP LAST NAME  XXXX

        EMP FIRST NAME XXXX

        EMP SOCIAL SEC XXXXXXXX

        EMP YEARS SRVC 00

        EMP LAST PROM  YYYY-MM-DD




F2 INQ   F3 EXIT   F4 ADD   F5 CHG   F6 DEL
```

Next go back to the main menu and choose the add option.[2]

```
EMPMMNU                    EMPLOYEE SUPPORT MENU                    EMNU

              ENTER THE NUMBER OF YOUR SELECTION,   THEN PRESS ENTER.


                     2    1. EMPLOYEE INQUIRY

                          2. EMPLOYEE ADD

                          3. EMPLOYEE CHANGE

                          4. EMPLOYEE DELETE










F3 EXIT
```

```
EMPMADD                      EMPLOYEE ADD                           EMAD


                     ENTER EMPLOYEE INFO, THEN PRESS PF4

         EMPLOYEE ID

         EMP LAST NAME

         EMP FIRST NAME

         EMP SOCIAL SEC

         EMP YEARS SRVC

         EMP LAST PROM




F2 INQ    F3 EXIT    F4 ADD    F5 CHG    F6 DEL
```

---

2 By now we can see it would be helpful to have an exit from each of the detail programs back to the main menu. I leave this as an enhancement for you so you can go through a maintenance cycle.

Now go back to the menu and choose the change option.

```
EMPMMNU                    EMPLOYEE SUPPORT MENU                    EMNU

             ENTER THE NUMBER OF YOUR SELECTION,   THEN PRESS ENTER.


                       3   1. EMPLOYEE INQUIRY

                           2. EMPLOYEE ADD

                           3. EMPLOYEE CHANGE

                           4. EMPLOYEE DELETE











  F3 EXIT
```

```
EMPMCHG                    EMPLOYEE CHANGE                          EMCH


        EMPLOYEE ->      ____        ENTER EMPLOYEE ID, THEN PRESS ENTER

        EMPLOYEE ID    XXXX

        EMP LAST NAME  XXXX

        EMP FIRST NAME XXXX

        EMP SOCIAL SEC XXXXXXXX

        EMP YEARS SRVC 00

        EMP LAST PROM  YYYY-MM-DD




  F2 INQ   F3 EXIT   F4 ADD   F5 CHG   F6 DEL
```

Finally choose the delete option.

```
EMPMMNU                    EMPLOYEE SUPPORT MENU                        EMNU

              ENTER THE NUMBER OF YOUR SELECTION,   THEN PRESS ENTER.


                       4    1. EMPLOYEE INQUIRY

                            2. EMPLOYEE ADD

                            3. EMPLOYEE CHANGE

                            4. EMPLOYEE DELETE









   F3 EXIT



EMPMDEL                      EMPLOYEE DELETE                            EMDE


        EMPLOYEE ->              ENTER EMPLOYEE ID, THEN PRESS ENTER

        EMPLOYEE ID    XXXX

        EMP LAST NAME  XXXX

        EMP FIRST NAME XXXX

        EMP SOCIAL SEC XXXXXXXXX

        EMP YEARS SRVC 00

        EMP LAST PROM  YYYY-MM-DD





   F2 INQ   F3 EXIT   F4 ADD   F5 CHG   F6 DEL
```

All looks good. Now on to the data operations.

EMPPGINQ
Ok let's start out with the primary display and entry after choosing from the main menu. Let's use 7777 as the employee id. Enter 7777 and press ENTER.

```
EMPMINQ                    EMPLOYEE INQUIRY                         EMIN

        EMPLOYEE ->     7777        ENTER EMPLOYEE ID, THEN PRESS ENTER

        EMPLOYEE ID    XXXX

        EMP LAST NAME  XXXX

        EMP FIRST NAME XXXX

        EMP SOCIAL SEC XXXXXXXXX

        EMP YEARS SRVC 00

        EMP LAST PROM  YYYY-MM-DD




     F2 INQ   F3 EXIT   F4 ADD   F5 CHG   F6 DEL



EMPMINQ                    EMPLOYEE INQUIRY                         EMIN

        EMPLOYEE ->     7777        ENTER EMPLOYEE ID, THEN PRESS ENTER

        EMPLOYEE ID    7777

        EMP LAST NAME  JACKSON

        EMP FIRST NAME JOSEPH

        EMP SOCIAL SEC 382746236

        EMP YEARS SRVC 17

        EMP LAST PROM  2017-01-01




     F2 INQ   F3 EXIT   F4 ADD   F5 CHG   F6 DEL
```

This looks good. Now let's try passing the same employee number from another program, such as the change program. Bring up employee 7777 on the change screen, and then press PF2 to transfer to the inquiry program.

```
EMPMCHG                     EMPLOYEE CHANGE                          EMCH


      EMPLOYEE ->    7777       ENTER EMPLOYEE ID, THEN PRESS ENTER

      EMPLOYEE ID    7777

      EMP LAST NAME  JACKSON

      EMP FIRST NAME JOSEPH

      EMP SOCIAL SEC 382746236

      EMP YEARS SRVC 17

      EMP LAST PROM  2017-01-01




MAKE CHANGES AND THEN PRESS PF5
F2 INQ   F3 EXIT   F4 ADD   F5 CHG   F6 DEL
```

```
EMPMINQ                     EMPLOYEE INQUIRY                         EMIN


      EMPLOYEE ->    7777       ENTER EMPLOYEE ID, THEN PRESS ENTER

      EMPLOYEE ID    7777

      EMP LAST NAME  JACKSON

      EMP FIRST NAME JOSEPH

      EMP SOCIAL SEC 382746236

      EMP YEARS SRVC 17

      EMP LAST PROM  2017-01-01




 F2 INQ   F3 EXIT   F4 ADD   F5 CHG   F6 DEL
```

Finally we should test the transfers from the inquiry screen to the add, change and delete screens. Start with inquiry and press PF4 to go to the add screen. Note that the employee id won't appear on the add screen because the add screen assumes we are going to enter a new number.

```
EMPMINQ                    EMPLOYEE INQUIRY                      EMIN


        EMPLOYEE ->    7777      ENTER EMPLOYEE ID, THEN PRESS ENTER

        EMPLOYEE ID    7777

        EMP LAST NAME  JACKSON

        EMP FIRST NAME JOSEPH

        EMP SOCIAL SEC 382746236

        EMP YEARS SRVC 17

        EMP LAST PROM  2017-01-01




    F2 INQ   F3 EXIT   F4 ADD   F5 CHG   F6 DEL

EMPMADD                     EMPLOYEE ADD                         EMAD

                   ENTER EMPLOYEE INFO, THEN PRESS PF4

        EMPLOYEE ID

        EMP LAST NAME

        EMP FIRST NAME

        EMP SOCIAL SEC

        EMP YEARS SRVC

        EMP LAST PROM




    ENTER DATA FOR NEW EMPLOYEE, THEN PRESS PF4 TO ADD
    F2 INQ   F3 EXIT   F4 ADD   F5 CHG   F6 DEL
```

Also go ahead and add an employee to ensure all attribute setting, switches and variables are correct.

```
EMPMADD                    EMPLOYEE ADD                              EMAD

                    ENTER EMPLOYEE INFO, THEN PRESS PF4

        EMPLOYEE ID    1111

        EMP LAST NAME  stone

        EMP FIRST NAME steven

        EMP SOCIAL SEC 385610088

        EMP YEARS SRVC 12

        EMP LAST PROM  2016-01-01




 ENTER DATA FOR NEW EMPLOYEE, THEN PRESS PF4 TO ADD
 F2 INQ   F3 EXIT   F4 ADD   F5 CHG   F6 DEL
```

```
EMPMADD                    EMPLOYEE ADD                              EMAD

                    ENTER EMPLOYEE INFO, THEN PRESS PF4

        EMPLOYEE ID    1111

        EMP LAST NAME  STONE

  .     EMP FIRST NAME STEVEN

        EMP SOCIAL SEC 385610088

        EMP YEARS SRVC 12

        EMP LAST PROM  2016-01-01




 EMPLOYEE ADDED SUCCESSFULLY
 F2 INQ   F3 EXIT   F4 ADD   F5 CHG   F6 DEL
```

All looks well, so let's move on to the add screen.

EMPPGADD

We just added a record, so let's try transferring to the inquiry screen, the change screen and then the delete screen. Press PF2.

```
EMPMADD                    EMPLOYEE ADD                         EMAD


                 ENTER EMPLOYEE INFO, THEN PRESS PF4

        EMPLOYEE ID    1111

        EMP LAST NAME  STONE

        EMP FIRST NAME STEVEN

        EMP SOCIAL SEC 385610088

        EMP YEARS SRVC 12

        EMP LAST PROM  2016-01-01
```

EMPLOYEE ADDED SUCCESSFULLY
```
F2 INQ   F3 EXIT   F4 ADD   F5 CHG   F6 DEL
```

```
EMPMINQ                   EMPLOYEE INQUIRY                     EMIN


        EMPLOYEE ->           ENTER EMPLOYEE ID, THEN PRESS ENTER

        EMPLOYEE ID    1111

        EMP LAST NAME  STONE

        EMP FIRST NAME STEVEN

        EMP SOCIAL SEC 385610088

        EMP YEARS SRVC 12

        EMP LAST PROM  2016-01-01




    F2 INQ   F3 EXIT   F4 ADD   F5 CHG   F6 DEL
```

Next let's add two more records, and in the process transfer to the change and delete screens, respectively.

```
EMPMADD                     EMPLOYEE ADD


                    ENTER EMPLOYEE INFO, THEN PRESS PF4

        EMPLOYEE ID    1212

        EMP LAST NAME  SAMPLE

        EMP FIRST NAME RECORD

        EMP SOCIAL SEC 373737373

        EMP YEARS SRVC 04

        EMP LAST PROM  2017-01-01




    EMPLOYEE ADDED SUCCESSFULLY
    F2 INQ   F3 EXIT   F4 ADD   F5 CHG   F6 DEL
```

Now press PF5.

```
EMPMCHG                  EMPLOYEE CHANGE                        EMCH


        EMPLOYEE ->              ENTER EMPLOYEE ID, THEN PRESS ENTER

        EMPLOYEE ID    1212

        EMP LAST NAME  SAMPLE

        EMP FIRST NAME RECORD

        EMP SOCIAL SEC 373737373

        EMP YEARS SRVC 04

        EMP LAST PROM  2017-01-01




    MAKE CHANGES AND THEN PRESS PF5
    F2 INQ   F3 EXIT   F4 ADD   F5 CHG   F6 DEL
```

Transfer back to the add screen and add the second record.

```
EMPMADD                        EMPLOYEE ADD                              EMAD


                    ENTER EMPLOYEE INFO, THEN PRESS PF4

         EMPLOYEE ID   2424

         EMP LAST NAME   TWO

         EMP FIRST NAME SAMPLE

         EMP SOCIAL SEC 747474747

         EMP YEARS SRVC 13

         EMP LAST PROM  2016-01-01




    EMPLOYEE ADDED SUCCESSFULLY
    F2 INQ    F3 EXIT    F4 ADD    F5 CHG    F6 DEL
```

Now press PF6.

```
EMPMDEL                        EMPLOYEE DELETE                           EMDE

         EMPLOYEE ->              ENTER EMPLOYEE ID, THEN PRESS ENTER

         EMPLOYEE ID   2424

         EMP LAST NAME   TWO

         EMP FIRST NAME SAMPLE

         EMP SOCIAL SEC 747474747

         EMP YEARS SRVC 13

         EMP LAST PROM  2016-01-01




    PRESS PF6 TO DELETE THIS RECORD
    F2 INQ    F3 EXIT    F4 ADD    F5 CHG    F6 DEL
```

All looks good with transferring from the add screen. Let's move on to the change screen.

EMPPGCHG

Let's bring up the 2424 record that we just created on the change screen. Now transfer to the inquiry screen, the delete screen, and finally the add screen (the latter will not process any transferred data except the program name).

```
EMPMCHG                      EMPLOYEE CHANGE                        EMCH


        EMPLOYEE ->    2424       ENTER EMPLOYEE ID, THEN PRESS ENTER

        EMPLOYEE ID    2424

        EMP LAST NAME  TWO

        EMP FIRST NAME SAMPLE

        EMP SOCIAL SEC 747474747

        EMP YEARS SRVC 13

        EMP LAST PROM  2016-01-01




   MAKE CHANGES AND THEN PRESS PF5
   F2 INQ   F3 EXIT   F4 ADD   F5 CHG   F6 DEL
```

Press PF2.

```
EMPMINQ                      EMPLOYEE INQUIRY                       EMIN


        EMPLOYEE ->    2424       ENTER EMPLOYEE ID, THEN PRESS ENTER

        EMPLOYEE ID    2424

        EMP LAST NAME  TWO

        EMP FIRST NAME SAMPLE

        EMP SOCIAL SEC 747474747

        EMP YEARS SRVC 13

        EMP LAST PROM  2016-01-01




   F2 INQ   F3 EXIT   F4 ADD   F5 CHG   F6 DEL
```

Transfer back to the change screen.

```
EMPMCHG                    EMPLOYEE CHANGE                         EMCH


        EMPLOYEE ->    2424       ENTER EMPLOYEE ID, THEN PRESS ENTER

        EMPLOYEE ID    2424

        EMP LAST NAME  TWO

        EMP FIRST NAME SAMPLE

        EMP SOCIAL SEC 747474747

        EMP YEARS SRVC 13

        EMP LAST PROM  2016-01-01




    MAKE CHANGES AND THEN PRESS PF5
    F2 INQ   F3 EXIT   F4 ADD   F5 CHG   F6 DEL
```

Now press PF6.

```
EMPMDEL                    EMPLOYEE DELETE                         EMDE


        EMPLOYEE ->    2424       ENTER EMPLOYEE ID, THEN PRESS ENTER

        EMPLOYEE ID    2424

        EMP LAST NAME  TWO

        EMP FIRST NAME SAMPLE

        EMP SOCIAL SEC 747474747

        EMP YEARS SRVC 13

        EMP LAST PROM  2016-01-01




    PRESS PF6 TO DELETE THIS RECORD
    F2 INQ   F3 EXIT   F4 ADD   F5 CHG   F6 DEL
```

Now go back to the change screen.

```
EMPMCHG                    EMPLOYEE CHANGE                          EMCH


        EMPLOYEE ->    2424      ENTER EMPLOYEE ID, THEN PRESS ENTER

        EMPLOYEE ID    2424

        EMP LAST NAME  TWO

        EMP FIRST NAME SAMPLE

        EMP SOCIAL SEC 747474747

        EMP YEARS SRVC 13

        EMP LAST PROM  2016-01-01




 MAKE CHANGES AND THEN PRESS PF5
 F2 INQ   F3 EXIT   F4 ADD   F5 CHG   F6 DEL
```

Press PF4.

```
EMPMADD                    EMPLOYEE ADD                             EMAD


                   ENTER EMPLOYEE INFO, THEN PRESS PF4

        EMPLOYEE ID

        EMP LAST NAME

        EMP FIRST NAME

        EMP SOCIAL SEC

        EMP YEARS SRVC

        EMP LAST PROM




 ENTER DATA FOR NEW EMPLOYEE, THEN PRESS PF4 TO ADD
 F2 INQ   F3 EXIT   F4 ADD   F5 CHG   F6 DEL
```

Again, the above is correct because the add screen should not process any passed employee number.

EMPPGDEL

Finally, let's test the delete screen. We can use employee 3333. Let's bring it up on the delete screen.

```
EMPMDEL                     EMPLOYEE DELETE                          EMDE


        EMPLOYEE ->    3333       ENTER EMPLOYEE ID, THEN PRESS ENTER

        EMPLOYEE ID    3333

        EMP LAST NAME  RADISSON

        EMP FIRST NAME BENTLEY

        EMP SOCIAL SEC 777777777

        EMP YEARS SRVC 46

        EMP LAST PROM  2015-07-01




  PRESS PF6 TO DELETE THIS RECORD
  F2 INQ   F3 EXIT   F4 ADD   F5 CHG   F6 DEL
```

Now transfer to the inquiry screen with PF2.

```
EMPMINQ                     EMPLOYEE INQUIRY                         EMIN


        EMPLOYEE ->    3333       ENTER EMPLOYEE ID, THEN PRESS ENTER

        EMPLOYEE ID    3333

        EMP LAST NAME  RADISSON

        EMP FIRST NAME BENTLEY

        EMP SOCIAL SEC 777777777

        EMP YEARS SRVC 46

        EMP LAST PROM  2015-07-01




  F2 INQ   F3 EXIT   F4 ADD   F5 CHG   F6 DEL
```

Now transfer back to the delete program

```
EMPMDEL                    EMPLOYEE DELETE                          EMDE


      EMPLOYEE ->    3333      ENTER EMPLOYEE ID, THEN PRESS ENTER

      EMPLOYEE ID    3333

      EMP LAST NAME  RADISSON

      EMP FIRST NAME BENTLEY

      EMP SOCIAL SEC 777777777

      EMP YEARS SRVC 46

      EMP LAST PROM  2015-07-01




  PRESS PF6 TO DELETE THIS RECORD
  F2 INQ   F3 EXIT   F4 ADD   F5 CHG   F6 DEL
```

And then transfer to the change program using PF5.

```
EMPMCHG                    EMPLOYEE CHANGE                          EMCH


      EMPLOYEE ->    3333      ENTER EMPLOYEE ID, THEN PRESS ENTER

      EMPLOYEE ID    3333

      EMP LAST NAME  RADISSON

      EMP FIRST NAME BENTLEY

      EMP SOCIAL SEC 777777777

      EMP YEARS SRVC 46

      EMP LAST PROM  2015-07-01




  MAKE CHANGES AND THEN PRESS PF5
  F2 INQ   F3 EXIT   F4 ADD   F5 CHG   F6 DEL
```

Now transfer back to the delete program

```
EMPMDEL                    EMPLOYEE DELETE                         EMDE


        EMPLOYEE ->    3333      ENTER EMPLOYEE ID, THEN PRESS ENTER

        EMPLOYEE ID    3333

        EMP LAST NAME  RADISSON

        EMP FIRST NAME BENTLEY

        EMP SOCIAL SEC 777777777

        EMP YEARS SRVC 46

        EMP LAST PROM  2015-07-01




   PRESS PF6 TO DELETE THIS RECORD
   F2 INQ   F3 EXIT   F4 ADD   F5 CHG   F6 DEL
```

Now transfer to the add program. Notice we do not pass the employee number to the add program since there is no need to add an already existing employee.

```
EMPMADD                    EMPLOYEE ADD                            EMAD

                   ENTER EMPLOYEE INFO, THEN PRESS PF4

        EMPLOYEE ID

        EMP LAST NAME

        EMP FIRST NAME

        EMP SOCIAL SEC

        EMP YEARS SRVC

        EMP LAST PROM




   ENTER DATA FOR NEW EMPLOYEE, THEN PRESS PF4 TO ADD
   F2 INQ   F3 EXIT   F4 ADD   F5 CHG   F6 DEL
```

That's it, looks like everything works. Our integration test is finished. This also completes our work on CICS with DB2. Now let's move on to CICS with VSAM.

134

Chapter Four: CICS With VSAM

VSAM Quick Review

Introduction
Virtual Storage Access Method (VSAM) is an IBM DASD (direct access storage device) file storage access method. It has been used for many years, including with the Multiple Virtual Storage (MVS) architecture and now in z/OS. VSAM offers four data set organizations:

1. Key Sequenced Data Set (KSDS)
2. Entry Sequenced Data Set (ESDS)
3. Relative Record Data Set (RRDS)
4. Linear Data Set (LDS)

The KSDS, RRDS and ESDS organizations all are record-based. The LDS organization uses a sequence of pages without a predefined record structure.

VSAM records are either fixed or variable length. Records are organized in fixed-size blocks called Control Intervals (CIs). The CI's are organized into larger structures called Control Areas (CAs). Control Interval sizes are measured in bytes — for example 4 kilobytes — while Control Area sizes are measured in disk tracks or cylinders. When a VSAM file is read, a complete Control Interval will be transferred to memory.

The Access Method Services utility program IDCAMS is used to define and delete VSAM data sets. In addition, you can write custom programs in COBOL, PLI and Assembler to access VSAM datasets using Data Definition (DD) statements in Job Control Language (JCL), or via dynamic allocation or in online regions such as in Customer Information Control System (CICS).

Types of VSAM Files

Key Sequence Data Set (KSDS)
This organization type is the most commonly used. Each record has one or more key fields and a record can be retrieved (or inserted) by key value. This provides random access to data. Records are of variable length.

Entry Sequence Data Set (ESDS)
This organization keeps records in the order in which they were entered. Records must be accessed sequentially. This organization is used by IMS for overflow datasets.

Relative Record Data Set (RRDS)
This organization is based on record retrieval number; record 1, record 2, etc. This provides random access but the program must have a way of knowing which record number it is looking for.

Linear Data Set (LDS)
This organization is a byte-stream data set. It is rarely used by application programs.

We'll focus on KSDS files because they are the most commonly used and the most useful. Also this is the kind of storage and retrieval method we need for our Employee Support application.

A KSDS cluster consists of following two components:

Index – the index component of the KSDS cluster is comprised of the list of key values for the records in the cluster with pointers to the corresponding records in the data component. The index component relates the key of each record to the record's relative location in the data set. When a record is added or deleted, this index is updated.

Data – the data component of the KSDS cluster contains the actual data. Each record in the data component of a KSDS cluster contains a key field with same number of characters and occurs in the same relative position in each record.

This is just a brief introduction to VSAM. If you need more background, check out **Quick Start Training for IBM z/OS Application Developers, Volume 2.** The first chapter gives you a more complete introduction to VSAM.

Creating the VSAM File
You create VSAM files using the IDCAMS utility. Here is sample JCL to create our employee file. We'll use 80 byte records which will be enough to accommodate all our field data. Also we'll define the key as the first 4 bytes of each record which is the employee id.

```
//USER01D JOB MSGLEVEL=(1,1),NOTIFY=&SYSUID
//*
//************************************************
//* DEFINE VSAM KSDS CLUSTER
//************************************************
//JS010    EXEC PGM=IDCAMS
//SYSUDUMP DD SYSOUT=*
//SYSPRINT DD SYSOUT=*
//SYSOUT   DD SYSOUT=*
//SYSIN    DD  *
  DEFINE CLUSTER(NAME(USER01.EMPLOYEE)   -
  RECSZ(80 80)       -
  TRK(2,1)           -
  FREESPACE(5,10)  -
  KEYS(4,0)          -
  CISZ(4096)         -
  VOLUMES(DEVHD1)  -
  INDEXED)           -
  INDEX(NAME(USER01.EMPLOYEE.INDEX)) -
  DATA(NAME(USER01.EMPLOYEE.DATA))
/*
//SYSPRINT DD SYSOUT=*
//SYSOUT   DD SYSOUT=*
//SYSUDUMP DD SYSOUT=*
//*
```

This creates a catalog entry with two datasets, one for data and one for the index.

```
DSLIST - Data Sets Matching USER01.EMPLOYEE                    Row 1 of 11
Command ===>                                          Scroll ===> CSR

Command - Enter "/" to select action             Message        Volume
-------------------------------------------------------------------------------
        USER01.EMPLOYEE                                          *VSAM*
        USER01.EMPLOYEE.DATA                                     DEVHD1
        USER01.EMPLOYEE.INDEX                                    DEVHD1
***************************** End of Data Set list ****************************
```

Loading and Unloading VSAM Files

You can add data to a VSAM KSDS in several ways:

1. Copying data from a flat file with the IDCAMS utility.

2. Using IBM's File Manager product.

3. Using an application program.

We'll show examples of loading data using a flat file with the IDCAMS utility, and also entering data with the File Manager utility. Here are the columns and data types for our table which we will name EMPLOYEE.

Field Name	Type
EMP_ID	Numeric 4 bytes
EMP_LAST_NAME	Character(30)
EMP_FIRST_NAME	Character(20)
EMP_SERVICE_YEARS	Numeric 2 bytes
EMP_PROMOTION_DATE	Date in format YYYY-MM-DD

Now let's say we have created a text data file in this format. We can browse it:

```
BROWSE    USER01.EMPLOYEE.LOAD                      Line 00000000 Col 001 080
 Command ===>                                            Scroll ===> CSR
----+----1----+----2----+----3----+----4----+----5----+----6----+----7----+----8
3217JOHNSON                   EDWARD              042017-01-01
7459STEWART                   BETTY               072016-07-31
9134FRANKLIN                  BRIANNA             032016-10-01
4720SCHULTZ                   TIM                 092017-01-01
6288WILLARD                   JOE                 062016-01-01
1122JENKINS                   DEBORAH             052016-09-01
```

Note however that before we load the VSAM file, we need to sort the input records in key sequence. Otherwise IDCAMS will give us an error. You can simply edit the file in ISPF and issue a SORT 1 4 command (space between 1 and 4) to sort the records. Here is the resulting sorted data. Now we are ready.

```
BROWSE    USER01.EMPLOYEE.LOAD                      Line 00000000 Col 001 080
 Command ===>                                            Scroll ===> CSR
----+----1----+----2----+----3----+----4----+----5----+----6----+----7----+----8
1122JENKINS                   DEBORAH             052016-09-01
3217JOHNSON                   EDWARD              042017-01-01
4720SCHULTZ                   TIM                 092017-01-01
6288WILLARD                   JOE                 062016-01-01
7459STEWART                   BETTY               072016-07-31
9134FRANKLIN                  BRIANNA             032016-10-01
```

VSAM Batch Updates with IDCAMS
We can use the following IDCAMS JCL to load the VSAM file.

```
//USER01D JOB 'WINGATE',MSGLEVEL=(1,1),NOTIFY=&SYSUID
//*
//* REPRO/COPY DATA FROM PS TO VSAM KSDS
//*
//JS010   EXEC PGM=IDCAMS
//SYSPRINT DD SYSOUT=*
//SYSOUT   DD SYSOUT=*
//SYSUDUMP DD SYSOUT=*
//SYSIN    DD *
  REPRO - INDATASET (USER01.EMPLOYEE.LOAD) -
  OUTDATASET(USER01.EMPLOYEE)
/*
//
```

Once loaded, we can view the data using the Browse function.

```
Browse            USER01.EMPLOYEE.DATA                    Top of 6
Command ===>                                              Scroll PAGE
                        Type DATA     RBA                 Format CHAR
                                          Col 1
----+----10---+----2----+----3----+----4----+----5----+----6----+----7----+----
1122JENKINS                    DEBORAH            052016-09-01
3217JOHNSON                    EDWARD             042017-01-01
4720SCHULTZ                    TIM                092017-01-01
6288WILLARD                    JOE                062016-01-01
7459STEWART                    BETTY              072016-07-31
9134FRANKLIN                   BRIANNA            032016-10-01
```

To edit the data you will need to use a tool such as File Manager. Let's do this next.

VSAM Updates with File Manager

You can perform adds, changes and deletes to data records in File Manager. First, it will be useful if we create a file layout to assist us with viewing and updating data. Let's create a COBOL layout as follows.

```
BROWSE    USER01.COPYLIB(EMPLOYEE) - 01.00       Line 00000000 Col 001 080
 Command ===>                                             Scroll ===> CSR
****************************************************************************
* COBOL DECLARATION FOR VSAM FILE EMPLOYEE                      *
****************************************************************************
 01   EMPLOYEE.
      05  EMP-ID              PIC 9(04).
      05  EMP-LAST-NAME       PIC X(30).
      05  EMP-FIRST-NAME      PIC X(20).
      05  EMP-SERVICE-YEARS   PIC 9(02).
      05  EMP-PROMOTION-DATE  PIC X(10).
      05  EMP-SSN             PIC X(09).
      05  FILLER              PIC X(05).
```

Now let's go to File Manager. Below is the main FM menu. Select the EDIT option.

```
File Manager                    Primary Option Menu
Command ===>

0   Settings        Set processing options              User ID . : USER01
1   View            View data                           System ID : MATE
2   Edit            Edit data                           Appl ID . : FMN
3   Utilities       Perform utility functions           Version . : 11.1.0
4   Tapes           Tape specific functions             Terminal. : 3278
5   Disk/VSAM       Disk track and VSAM CI functions    Screen. . : 2
6   OAM             Work with OAM objects               Date. . . : 2018/03/07
7   Templates       Template and copybook utilities     Time. . . : 02:41
8   HFS             Access Hierarchical File System
9   WebSphere MQ    List, view and edit MQ data
X   Exit            Terminate File Manager
```

Enter your file name, copybook file name, and select the processing option 1.

```
File Manager                    Edit Entry Panel
Command ===>

Input Partitioned, Sequential or VSAM Data Set, or HFS file:
    Data set/path name 'USER01.EMPLOYEE'                              +
    Member . . . . . .              (Blank or pattern for member list)
    Volume serial  . .              (If not cataloged)
    Start position . .                             +
    Record limit . . .          Record sampling
    Inplace edit . . .          (Prevent inserts and deletes)
Copybook or Template:
    Data set name  . . 'USER01.COPYLIB(EMPLOYEE)'
    Member . . . . . .              (Blank or pattern for member list)
Processing Options:
  Copybook/template   Start position type   Enter "/" to select option
  1  1. Above           1. Key                Edit template    Type (1,2,S)
     2. Previous        2. RBA                Include only selected records
     3. None            3. Record number      Binary mode, reclen 80
     4. Create dynamic  4. Formatted key      Create audit trail
```

Now you will see this screen. Notice the format is TABL which shows the data in list format. If you want to change it to show one record at a time, type over the TABL with SNGL (which means single record).

```
Edit              USER01.EMPLOYEE                        Top of 6
Command ===>                                             Scroll PAGE
     Key                      Type KSDS     RBA          Format TABL
        EMP-ID EMP-LAST-NAME              EMP-FIRST-NAME  EMP-SERVICE-
              #2 #3                       #4                        #5
           ZD 1:4 AN 5:30                 AN 35:20        ZD 55:2
           <--->  <---+----1----+----2----+----> <---+----1----+---->  <->
****** ****  Top of data   ****
000001    1122 JENKINS                   DEBORAH                    5
000002    3217 JOHNSON                   EDWARD                     4
000003    4720 SCHULTZ                   TIM                        9
000004    6288 WILLARD                   JOE                        6
000005    7459 STEWART                   BETTY                      7
000006    9134 FRANKLIN                  BRIANNA                    3
****** ****  End of data   ****
```

Now you can edit each field on the record except the key. You cannot change the key, although you can specify a different key to bring up a different record. Let's bring up employee 6288.

```
Edit              USER01.EMPLOYEE                        Rec 1 of 6
Command ===>                                             Scroll PAGE
Key 1122                     Type KSDS     RBA 0         Format SNGL
                                                  Top Line is 1    of 6
Current 01: EMPLOYEE                              Length 80
Field               Data
EMP-ID                 1122
EMP-LAST-NAME       JENKINS
EMP-FIRST-NAME      DEBORAH
EMP-SERVICE-YEARS      5
EMP-PROMOTION-DATE  2016-09-01
FILLER
***  End of record   ***
```

Now we can change this record. Let's change the years of service to 8.

```
Edit              USER01.EMPLOYEE                        Rec 4 of 6
Command ===>                                             Scroll PAGE
Key 6288                     Type KSDS     RBA 240       Format SNGL
                                                  Top Line is 1    of 6
Current 01: EMPLOYEE                              Length 80
Field               Data
EMP-ID                 6288
EMP-LAST-NAME       WILLARD
EMP-FIRST-NAME      JOE
EMP-SERVICE-YEARS      6
EMP-PROMOTION-DATE  2016-01-01
FILLER
```

```
***  End of record  ***
```

Now you can type SAVE on the command line or simply PF3 to exit from the record.

```
File Manager                 Edit Entry Panel              1 record(s) updated
Command ===>

Input Partitioned, Sequential or VSAM Data Set, or HFS file:
   Data set/path name  'USER01.EMPLOYEE'                              +
   Member . . . . . .              (Blank or pattern for member list)
   Volume serial  . .              (If not cataloged)
   Start position . .                                   +
   Record limit . . .          Record sampling
   Inplace edit . . .              (Prevent inserts and deletes)
Copybook or Template:
   Data set name  . .  'USER01.COPYLIB(EMPLOYEE)'
   Member . . . . . .              (Blank or pattern for member list)
Processing Options:
 Copybook/template    Start position type    Enter "/" to select option
 1  1. Above            1. Key                  Edit template    Type (1,2,S)
    2. Previous         2. RBA                  Include only selected records
    3. None             3. Record number        Binary mode, reclen 80
    4. Create dynamic   4. Formatted key        Create audit trail
```

Now let's see how we can insert and delete records. Actually it is pretty simple. If you are in table mode, you just use the I line command to insert a record, or the D line command to delete one. Let's add a record for employee 1111 who is Sandra Smith with 9 years of service and a promotion date of 01/01/2017. To do this, type I on the first line of detail.

```
Edit              USER01.EMPLOYEE                          Rec 1 of 6
Command ===>                                               Scroll PAGE
     Key 1122              Type KSDS      RBA 0            Format TABL
        EMP-ID EMP-LAST-NAME                EMP-FIRST-NAME  EMP-SERVICE-
           #2 #3                            #4                      #5
           ZD 1:4 AN 5:30                   AN 35:20        ZD 55:2
           <--->  <---+----1----+----2----+----> <---+----1----+----> <->
I00001     1122 JENKINS                     DEBORAH               5
000002     3217 JOHNSON                     EDWARD                4
000003     4720 SCHULTZ                     TIM                   9
000004     6288 WILLARD                     JOE                   8
000005     7459 STEWART                     BETTY                 7
000006     9134 FRANKLIN                    BRIANNA               3
****** ****  End of data   ****
```

Now you can enter the data. You will need to scroll to the right to add the correct years of service and promotion date. When you finish press ENTER and you'll see this screen.

```
Edit                USER01.EMPLOYEE                      Rec 1 of 7
Command ===>                                             Scroll PAGE
      Key 1122                Type KSDS    RBA 0          Format TABL
        EMP-ID EMP-LAST-NAME                EMP-FIRST-NAME    EMP-SERVICE-
          #2 #3                             #4                         #5
        ZD 1:4 AN 5:30                      AN 35:20            ZD 55:2
          <---> <---+----1----+----2----+----> <---+----1----+----> <->
000001    1122 JENKINS                     DEBORAH                     5
000002    1111 SMITH                       SANDRA                      0
000003    3217 JOHNSON                     EDWARD                      4
000004    4720 SCHULTZ                     TIM                         9
000005    6288 WILLARD                     JOE                         8
000006    7459 STEWART                     BETTY                       7
000007    9134 FRANKLIN                    BRIANNA                     3
****** ****   End of data   ****
```

You could also switch to SNGL mode to make it easier to enter the data on one page. Let's do this.

```
Edit                USER01.EMPLOYEE                      Rec 1 of 7
Command ===>                                             Scroll PAGE
Key 1111                    Type KSDS    RBA 0           Format SNGL
                                             Top Line is 1    of 6
Current 01: EMPLOYEE                                     Length 80
Field                   Data
EMP-ID                     1111
EMP-LAST-NAME              SMITH
EMP-FIRST-NAME            SANDRA
EMP-SERVICE-YEARS           9
EMP-PROMOTION-DATE      2017-01-01
FILLER
***  End of record   ***
```

Now type SAVE on the command line.

```
Edit                USER01.EMPLOYEE                      Rec 1 of 7
Command ===>         SAVE                                Scroll PAGE
Key 1111                    Type KSDS    RBA 0           Format SNGL
                                             Top Line is 1    of 6
Current 01: EMPLOYEE                                     Length 80
Field                   Data
EMP-ID                     1111
EMP-LAST-NAME              SMITH
EMP-FIRST-NAME            SANDRA
EMP-SERVICE-YEARS           9
EMP-PROMOTION-DATE      2017-01-01
FILLER
***  End of record   ***
```

When you press Enter you can verify the record was saved.

```
Edit              USER01.EMPLOYEE                      1 record(s) updated
Command ===>                                                    Scroll PAGE
Key 1111                    Type KSDS      RBA 0               Format SNGL
                                               Top Line is 1    of 6
Current 01: EMPLOYEE                                         Length 80
Field            Data
EMP-ID             1111
EMP-LAST-NAME      SMITH
EMP-FIRST-NAME     SANDRA
EMP-SERVICE-YEARS     9
EMP-PROMOTION-DATE 2017-01-01
FILLER
***   End of record   ***
```

Finally, to delete a record, just go to TABL mode, find the record you want to delete, and use a D line action. Let's delete the record we just added.

```
Edit              USER01.EMPLOYEE                          Rec 1 of 7
Command ===>                                                 Scroll PAGE
      Key 1111                Type KSDS      RBA 0           Format TABL
        EMP-ID EMP-LAST-NAME                 EMP-FIRST-NAME  EMP-SERVICE-
          #2 #3                              #4                      #5
          ZD 1:4 AN 5:30                     AN 35:20           ZD 55:2
          <---> <---+----1----+----2----+----> <---+----1----+---->   <->
D00001     1111 SMITH                         SANDRA                    9
000002     1122 JENKINS                       DEBORAH                   5
000003     3217 JOHNSON                       EDWARD                    4
000004     4720 SCHULTZ                       TIM                       9
000005     6288 WILLARD                       JOE                       8
000006     7459 STEWART                       BETTY                     7
000007     9134 FRANKLIN                      BRIANNA                   3
****** ****   End of data   ****
```

The record has disappeared from the display.

```
Edit              USER01.EMPLOYEE                          Rec 1 of 6
Command ===>                                                 Scroll PAGE
      Key 1122                Type KSDS      RBA 80          Format TABL
        EMP-ID EMP-LAST-NAME                 EMP-FIRST-NAME  EMP-SERVICE-
          #2 #3                              #4                      #5
          ZD 1:4 AN 5:30                     AN 35:20           ZD 55:2
          <---> <---+----1----+----2----+----> <---+----1----+---->   <->
000001     1122 JENKINS                       DEBORAH                   5
000002     3217 JOHNSON                       EDWARD                    4
000003     4720 SCHULTZ                       TIM                       9
000004     6288 WILLARD                       JOE                       8
000005     7459 STEWART                       BETTY                     7
000006     9134 FRANKLIN                      BRIANNA                   3
****** ****   End of data   ****
```

You can either type SAVE on the command line, or simply exit the file and the delete action will be saved.

```
Edit                USER01.EMPLOYEE                        1 record(s) updated
Command ===>                                                 Scroll PAGE
      Key 1122                    Type KSDS    RBA 80         Format TABL
          EMP-ID EMP-LAST-NAME                 EMP-FIRST-NAME    EMP-SERVICE-
             #2 #3                             #4                         #5
             ZD 1:4 AN 5:30                    AN 35:20             ZD 55:2
             <---> <---+----1----+----2----+----> <---+----1----+---->       <->
000001    1122 JENKINS                        DEBORAH                       5
000002    3217 JOHNSON                        EDWARD                        4
000003    4720 SCHULTZ                        TIM                           9
000004    6288 WILLARD                        JOE                           8
000005    7459 STEWART                        BETTY                         7
000006    9134 FRANKLIN                       BRIANNA                       3
****** ****   End of data   ****
```

Now that we have some data, let's do some programming.

CICS Application Programming with VSAM

As we introduce VSAM as our data store, our application design does not change. All the screens will look exactly the same. The program structure and most of the logic is the same. The only thing that will change is the data store and the commands we use to access and update the data.

Record Structure

We introduced our COBOL copybook earlier. To make the fewest changes to our programs, we must make sure that our field names match what's already in the program. We'll only change one thing. We used fixed length records in our VSAM file, so we don't need the length and text fields for the last name and first name. Here's the layout.

```
01   DCLEMPLOYEE.
     05 EMP-ID                PIC 9(04).
     05 EMP-LAST-NAME         PIC X(30).
     05 EMP-FIRST-NAME        PIC X(20).
     05 EMP-SERVICE-YEARS     PIC 9(02).
     05 EMP-PROMOTION-DATE    PIC X(10).
     05 EMP-SSN               PIC X(09).
     05 FILLER                PIC X(05).
```

How to Define the VSAM File to CICS

Now we need to define our VSAM file to CICS. Invoke the CEDA transaction and type `DEF FILE`. We'll give the VSAM file a CICS name of `EMPLOYEE` and we must also specify the actual z/OS file name.

```
CEDA DEF FILE
 OVERTYPE TO MODIFY                                CICS RELEASE = 0670
  CEDA  DEFine File(          )
   File           ==> EMPLOYEE
   Group          ==> USER01
   DEScription    ==> EMPLOYEE SUPPORT FILE
  VSAM PARAMETERS
   DSNAme         ==> USER01.EMPLOYEE
   Password       ==>                   PASSWORD NOT SPECIFIED
   RLsaccess      ==> No                Yes | No
   LSRPOOLId       : 1                  1-8 | None
   LSRPOOLNum     ==> 001               1-255 | None
   READInteg      ==> Uncommitted       Uncommitted | Consistent | Repeatable
   DSNSharing     ==> Allreqs           Allreqs | Modifyreqs
   STRings        ==> 001               1-255
   Nsrgroup       ==>
  REMOTE ATTRIBUTES
   REMOTESystem ==>
   REMOTEName   ==>
 + REMOTE AND CFDATATABLE PARAMETERS
                                         SYSID=CICS APPLID=CICSTS42

 PF 1 HELP 2 COM 3 END         6 CRSR 7 SBH 8 SFH 9 MSG 10 SB 11 SF 12 CNCL
```

Press Enter and you should get a Define Successful message at the bottom.

```
OVERTYPE TO MODIFY                                      CICS RELEASE = 0670
  CEDA  DEFine File( EMPLOYEE )
   File             : EMPLOYEE
   Group            : USER01
   DEScription  ==> EMPLOYEE SUPPORT FILE
  VSAM PARAMETERS
   DSNAme       ==> USER01.EMPLOYEE
   Password     ==>                    PASSWORD NOT SPECIFIED
   RLsaccess    ==> No                 Yes | No
   LSRPOOLId     : 1                    1-8 | None
   LSRPOOLNum   ==> 001                 1-255 | None
   READInteg    ==> Uncommitted         Uncommitted | Consistent | Repeatable
   DSNSharing   ==> Allreqs             Allreqs | Modifyreqs
   STRings      ==> 001                 1-255
   Nsrgroup     ==>
  REMOTE ATTRIBUTES
   REMOTESystem ==>
   REMOTEName   ==>
+ REMOTE AND CFDATATABLE PARAMETERS

                                        SYSID=CICS APPLID=CICSTS42
   DEFINE SUCCESSFUL                    TIME: 02.52.08  DATE: 12/05/18
 PF 1 HELP 2 COM 3 END           6 CRSR 7 SBH 8 SFH 9 MSG 10 SB 11 SF 12 CNCL
```

Next, scroll forward a couple of pages until you see this screen. I suggest you specify YES on each operation. This means you can add, update, delete, read and browse the file.

```
OVERTYPE TO MODIFY OR PRESS ENTER TO EXECUTE            CICS RELEASE = 0670
  CEDA  DEFine File( EMPLOYEE )
+ DATA FORMAT
   RECORDFormat ==> V                  V | F
  OPERATIONS
   Add          ==> yes                No | Yes
   BRowse       ==> yes                No | Yes
   DELete       ==> yes                No | Yes
   READ         ==> yes                Yes | No
   UPDATE       ==> yes                No | Yes
  AUTO JOURNALLING
   JOurnal      ==> No                 No | 1-99
   JNLRead      ==> None               None | Updateonly | Readonly | All
   JNLSYNCRead  ==> No                 No | Yes
   JNLUpdate    ==> No                 No | Yes
   JNLAdd       ==> None               None | Before | AFter | ALl
   JNLSYNCWrite ==> Yes                Yes | No
  RECOVERY PARAMETERS
+ RECOVery     ==> None               None | Backoutonly | All

                                        SYSID=CICS APPLID=CICSTS42

 PF 1 HELP 2 COM 3 END           6 CRSR 7 SBH 8 SFH 9 MSG 10 SB 11 SF 12 CNCL
```

Next, we must install the file with this command:

```
CEDA INSTALL FILE(EMPLOYEE) GROUP(USER01)
```

Finally, you must open the file using the following CEMT command:

```
CEMT SET FILE(EMPLOYEE) OPEN
```

CICS Commands to Access the VSAM File
Now let's look at the CICS commands that are necessary to carry out file operations on our VSAM file.

Reading a Record
To perform direct access on the EMPLOYEE file, we issue the following CICS command:

```
EXEC CICS
READ
FILE('EMPLOYEE')
INTO (DCLEMPLOYEE)
RIDFLD(EMP-ID)
EQUAL
RESP(RESPONSE-CODE)
END-EXEC.
```

We will also need to check the status of the read. We can do this using the RESPONSE-CODE variable that we'll add to the program. This will replace our check of the SQLCODE that we used in the DB2 version of the program.

```
IF DFHRESP(NORMAL)
   <Do normal processing>
ELSE
   <Do error processing>
END-IF
```

If the return code is CICS-NORMAL then it means we got a good return from the action.

Adding a Record
You add a record by loading the employee record structure and then do a WRITE.

```
EXEC CICS
WRITE
FILE('EMPLOYEE')
FROM (DCLEMPLOYEE)
RESP(RESPONSE-CODE)
END-EXEC.
```

Updating a Record

Updating a record in CICS is two step. It involves reading the record with a lock on it, and then performing a rewrite command. To lock the record we include the keyword UPDATE in the READ command. Here are both the READ and REWRITE commands:

```
EXEC CICS
READ
FILE ('EMPLOYEE')
INTO (DCLEMPLOYEE)
RIDFLD (EMP-ID)
UPDATE
EQUAL
RESP (RESPONSE-CODE)

END-EXEC.

EXEC CICS
REWRITE
FILE ('EMPLOYEE')
FROM (DCLEMPLOYEE)
RESP (RESPONSE-CODE)
END-EXEC.
```

Deleting a Record

Deleting a record is fairly straightforward. You do not need to first retrieve the record. Simply load the key value into the record structure and issue the delete as follows:

```
EXEC CICS READ
DELETE
FILE ('EMPLOYEE')
RIDFLD (EMP-ID)
RESP (RESPONSE-CODE)
END-EXEC.
```

Revised Employee Support Programs

EMPPGINQ

We will make the following revisions to our inquiry program:

1. Remove DB2 include commands.
2. Add the new EMPLOYEE record structure.
3. Remove the -TEXT suffixes from EMP-LAST-NAME and EMP-FIRST-NAME fields.
4. Remove the DB2 query and replace with CICS READ.
5. Check for success using the CICS variables for good read (and missing record).

Now we'll make revisions to our employee support programs to use VSAM instead of DB2. Here is our revised code for the Employee Inquiry program:

```
       IDENTIFICATION DIVISION.
       PROGRAM-ID. EMPPGINQ.
      *****************************************************
      *  COBOL/CICS/DB2 PROGRAM TO DISPLAY AN EMPLOYEE  *
      *                                                 *
      *  AUTHOR        : ROBERT WINGATE                 *
      *  DATE-WRITTEN  : 2018-07-19                     *
      *****************************************************
       ENVIRONMENT DIVISION.
       DATA DIVISION.
       WORKING-STORAGE SECTION.
       01 WS-EMPNO        PIC 9(4).
       01 WS-EMP-SRV-YRS  PIC 9(2).
       01 WS-COMMAREA.
          05 WS-EMP-PASS   PIC 9(04) VALUE ZERO.
          05 WS-PGM-PASS   PIC X(08) VALUE SPACES.
          05 FILLER        PIC X(08).

       01 RESPONSE-CODE    PIC S9(08) COMP.
       01 RESPONSE-DISPLAY PIC S9(08) USAGE DISPLAY.

       01 PROGRAM-NAME     PIC X(08) VALUE SPACES.
       01  DCLEMPLOYEE.
           05 EMP-ID             PIC 9(04).
           05 EMP-LAST-NAME      PIC X(30).
           05 EMP-FIRST-NAME     PIC X(20).
           05 EMP-SERVICE-YEARS  PIC 9(02).
           05 EMP-PROMOTION-DATE PIC X(10).
           05 EMP-SSN            PIC X(09).
           05 FILLER             PIC X(05).

       01 SW-PASSED-DATA-SWITCH   PIC X(1) VALUE 'N'.
          88  SW-PASSED-DATA                VALUE 'Y'.
          88  SW-NO-PASSED-DATA             VALUE 'N'.

          COPY EMPMINQ.
          COPY DFHAID.
          COPY DFHBMSCA.

       LINKAGE SECTION.

       01 DFHCOMMAREA          PIC X(20).

       PROCEDURE DIVISION.

          SET SW-NO-PASSED-DATA TO TRUE

          IF EIBCALEN > ZERO
            MOVE DFHCOMMAREA  TO WS-COMMAREA
          END-IF.

          EVALUATE TRUE

            WHEN EIBCALEN = ZERO
```

```
                  MOVE LOW-VALUES    TO  EMPINQO
                  PERFORM SEND-MAP

              WHEN EIBAID = DFHCLEAR
                  MOVE LOW-VALUES    TO  EMPINQO
                  PERFORM SEND-MAP

              WHEN EIBAID = DFHPA1 OR DFHPA2 OR DFHPA3
                  CONTINUE

              WHEN EIBAID = DFHPF2
                  IF WS-PGM-PASS NOT EQUAL "EMPPGINQ"
                     SET SW-PASSED-DATA TO TRUE
                  END-IF
                  PERFORM PROCESS-PARA

              WHEN EIBAID = DFHPF3
                  MOVE LOW-VALUES TO  EMPINQO
                  MOVE "BYE, PRESS CLEAR KEY TO ENTER A TRANSACTION ID"
                       TO MESSAGEO
                  PERFORM SEND-MAP-DATA

                  EXEC CICS
                     RETURN
                  END-EXEC

              WHEN EIBAID = DFHPF4
                  MOVE 'EMPPGADD' TO PROGRAM-NAME
                  PERFORM BRANCH-TO-PROGRAM

                  EXEC CICS
                     RETURN
                  END-EXEC

              WHEN EIBAID = DFHPF5
                  MOVE 'EMPPGCHG' TO PROGRAM-NAME
                  PERFORM BRANCH-TO-PROGRAM

                  EXEC CICS
                     RETURN
                  END-EXEC

              WHEN EIBAID = DFHPF6
                  MOVE 'EMPPGDEL' TO PROGRAM-NAME
                  PERFORM BRANCH-TO-PROGRAM

                  EXEC CICS
                     RETURN
                  END-EXEC

              WHEN EIBAID = DFHENTER
                  PERFORM PROCESS-PARA

              WHEN OTHER
                  MOVE LOW-VALUES TO EMPINQO
                  MOVE "INVALID KEY PRESSED" TO MESSAGEO
                  PERFORM SEND-MAP-DATA
```

```
        END-EVALUATE.

        EXEC CICS
            RETURN TRANSID('EMIN')
            COMMAREA (WS-COMMAREA)
            LENGTH(20)
        END-EXEC.

    PROCESS-PARA.

        PERFORM RECEIVE-MAP.
        INITIALIZE DCLEMPLOYEE MESSAGEO
        IF SW-PASSED-DATA
            MOVE WS-EMP-PASS TO EMP-ID
        ELSE
            MOVE EMPINI    TO WS-EMPNO
            MOVE WS-EMPNO  TO EMP-ID
        END-IF

        EXEC CICS
            READ
            FILE('EMPLOYEE')
            INTO(DCLEMPLOYEE)
            RIDFLD(EMP-ID)
            EQUAL
            RESP(RESPONSE-CODE)
        END-EXEC.

        EVALUATE RESPONSE-CODE
            WHEN DFHRESP(NORMAL)
                MOVE EMP-ID                TO  WS-EMPNO WS-EMP-PASS
                MOVE WS-EMPNO              TO  EMPNOO
                MOVE EMP-LAST-NAME        TO LNAMEO
                MOVE EMP-FIRST-NAME       TO FNAMEO
                MOVE EMP-SSN              TO SOCSECO
                MOVE EMP-SERVICE-YEARS    TO WS-EMP-SRV-YRS
                MOVE WS-EMP-SRV-YRS       TO YRSSVCO
                MOVE EMP-PROMOTION-DATE   TO LSTPRMO

            WHEN DFHRESP(NOTFND)
                STRING "EMPLOYEE ID " DELIMITED BY SIZE
                WS-EMPNO DELIMITED BY SPACE
                " NOT FOUND" DELIMITED BY SIZE INTO MESSAGEO
                MOVE SPACES        TO EMPNOO
                MOVE SPACES        TO LNAMEO
                MOVE SPACES        TO FNAMEO
                MOVE SPACES        TO SOCSECO
                MOVE SPACES        TO YRSSVCO
                MOVE SPACES        TO LSTPRMO

            WHEN OTHER
                MOVE RESPONSE-CODE TO RESPONSE-DISPLAY
                STRING "UNKNOWN ERROR CODE: " DELIMITED BY SIZE
                    RESPONSE-DISPLAY DELIMITED BY SIZE
                  INTO MESSAGEO

        END-EVALUATE.
```

```
        MOVE DFHBMFSE TO EMPINF
        MOVE -1 TO EMPINL
        MOVE "EMPPGINQ" TO WS-PGM-PASS

        IF SW-PASSED-DATA
           PERFORM SEND-MAP
        ELSE
           PERFORM SEND-MAP-DATA
        END-IF.

    BRANCH-TO-PROGRAM.
        EXEC CICS
           XCTL PROGRAM(PROGRAM-NAME)
           COMMAREA (WS-COMMAREA)
           LENGTH(20)
        END-EXEC

        MOVE 'PROGRAM NOT AVAILABLE' TO MESSAGEO.
        MOVE DFHBMFSE TO EMPINF
        MOVE -1 TO EMPINL
        PERFORM SEND-MAP-DATA.

    SEND-MAP.
        EXEC CICS SEND
           MAP     ('EMPINQ')
           MAPSET ('EMPMINQ')
           FROM    (EMPINQO)
           ERASE
        END-EXEC.

    SEND-MAP-DATA.
        EXEC CICS SEND
           MAP     ('EMPINQ')
           MAPSET ('EMPMINQ')
           FROM    (EMPINQO)
           DATAONLY
        END-EXEC.

    RECEIVE-MAP.
         EXEC CICS RECEIVE
           MAP     ('EMPINQ')
           MAPSET ('EMPMINQ')
           INTO    (EMPINQI)
         END-EXEC.
```

Now we need to do a straight CICS compile and link (no DB2 bind). Here is the JCL I use for that.

```
//USER01D JOB MSGLEVEL=(1,1),NOTIFY=&SYSUID
//*
//*   COMPILE A COBOL + CICS PROGRAM
//*
//CICSCOB  EXEC CICSCOBC,
//            COPYLIB=USER01.COPYLIB,        <= COPYBOOK LIBRARY
//            SRCLIB=USER01.CICS.SRCLIB,     <= SOURCE LIBRARY
//            MEMBER=EMPPGINQ                 <= SOURCE MEMBER
```

Next we need to go into CICS and refresh the program load module using CEMT:

```
CEMT SET PROGRAM(EMPPGINQ) NEWCOPY
```

Finally we are ready to test. Let's bring up the EMIN screen and then display data for employee 3217:

```
EMPMINQ                    EMPLOYEE INQUIRY                      EMIN


       EMPLOYEE ->     ____      ENTER EMPLOYEE ID, THEN PRESS ENTER

       EMPLOYEE ID    XXXX

       EMP LAST NAME  XXXX

       EMP FIRST NAME XXXX

       EMP SOCIAL SEC XXXXXXXXX

       EMP YEARS SRVC 00

       EMP LAST PROM  YYYY-MM-DD




  F2 INQ   F3 EXIT   F4 ADD   F5 CHG   F6 DEL
```

Enter 3217, press ENTER and here is the result:

```
EMPMINQ                    EMPLOYEE INQUIRY                      EMIN


       EMPLOYEE ->     3217      ENTER EMPLOYEE ID, THEN PRESS ENTER

       EMPLOYEE ID    3217

       EMP LAST NAME  JOHNSON

       EMP FIRST NAME EDWARD

       EMP SOCIAL SEC 493082938

       EMP YEARS SRVC 04

       EMP LAST PROM  2017-01-01




  F2 INQ   F3 EXIT   F4 ADD   F5 CHG   F6 DEL
```

Great, it works! If your version does not work, please make sure to check each step and correct if necessary.

Now, to finish the DB2-to-VSAM conversion we'll need to do regression testing. So run all the unit test cases to make sure they still work. I've done mine and didn't find any errors.

Now move on to the add program!

EMPPGADD

For the add program we'll do these same conversions as for the inquiry except of course we will be writing a record instead of reading it. Make these changes:

1. Remove DB2 include commands
2. Add the new EMPLOYEE record structure
3. Remove the -TEXT suffixes from EMP-LAST-NAME and EMP-FIRST-NAME fields
4. Remove any code that attempts to calculate the length of the employee last name or first name fields
5. Remove the DB2 INSERT query and replace with CICS WRITE
6. Check for success using the CICS variables for successful transaction as well as missing record.

Give this a try, then take a break and come back and we'll compare code.
…..

Ok, I'm back. Here's my code for the add program:

```
IDENTIFICATION DIVISION.
PROGRAM-ID. EMPPGADD.

**************************************************
* COBOL/CICS/VSAM PROGRAM TO ADD AN EMPLOYEE    *
*                                               *
* AUTHOR        : ROBERT WINGATE                *
* DATE-WRITTEN  : 2018-07-21                    *
**************************************************

ENVIRONMENT DIVISION.

DATA DIVISION.

WORKING-STORAGE SECTION.
01 WS-EMPNO        PIC 9(04).
01 WS-EMP-SRV-YRS  PIC 9(02).
01 WS-COMMAREA.
```

```
    05 WS-EMP-PASS    PIC 9(04).
    05 WS-PGM-PASS    PIC X(08).
    05 FILLER         PIC X(08).

01 PROGRAM-NAME      PIC X(08) VALUE SPACES.

01 RESPONSE-CODE     PIC S9(08) VALUE 0.
01 RESPONSE-DISPLAY  PIC  9(08) VALUE 0.

01 RESPONSE-CODE2    PIC S9(08) VALUE 0.
01 RESPONSE-DISPLA2  PIC  9(08) VALUE 0.

01  DCLEMPLOYEE.
    05 EMP-ID               PIC 9(04).
    05 EMP-LAST-NAME        PIC X(30).
    05 EMP-FIRST-NAME       PIC X(20).
    05 EMP-SERVICE-YEARS    PIC 9(02).
    05 EMP-PROMOTION-DATE   PIC X(10).
    05 EMP-SSN              PIC X(09).
    05 FILLER               PIC X(05).

    COPY EMPMADD.
    COPY DFHAID.
    COPY DFHBMSCA.

LINKAGE SECTION.

01 DFHCOMMAREA        PIC X(20).

PROCEDURE DIVISION.

    IF EIBCALEN > ZERO
      MOVE DFHCOMMAREA  TO WS-COMMAREA
    END-IF.

    EVALUATE TRUE

      WHEN EIBCALEN = ZERO
        MOVE LOW-VALUES   TO  EMPADDO
        MOVE -1 TO EMPNOL
        PERFORM SEND-MAP

      WHEN EIBAID = DFHCLEAR
        MOVE LOW-VALUES   TO  EMPADDO
        MOVE -1 TO EMPNOL
        PERFORM SEND-MAP

      WHEN EIBAID = DFHPA1 OR DFHPA2 OR DFHPA3
        CONTINUE

      WHEN EIBAID = DFHPF2
        MOVE 'EMPPGINQ' TO PROGRAM-NAME
        PERFORM BRANCH-TO-PROGRAM

        EXEC CICS
          RETURN
        END-EXEC
```

156

```
         WHEN EIBAID = DFHPF3
           MOVE LOW-VALUES TO  EMPADDO
           MOVE -1 TO EMPNOL
           MOVE "BYE, PRESS CLEAR KEY TO ENTER A TRANSACTION ID"
              TO MESSAGEO
           PERFORM SEND-MAP-DATA

           EXEC CICS
             RETURN
           END-EXEC

         WHEN EIBAID = DFHPF4
*           PERFORM THE EDITS AND VALIDATIONS
*           IF NO ERRORS THEN INSERT THE RECORDS

           IF WS-PGM-PASS NOT EQUAL "EMPPGADD"
              MOVE LOW-VALUES   TO  EMPADDO
              MOVE -1 TO EMPNOL
              MOVE
              "ENTER DATA FOR NEW EMPLOYEE, THEN PRESS PF4 TO ADD"
              TO MESSAGEO
              MOVE "EMPPGADD" TO WS-PGM-PASS
              PERFORM SEND-MAP
           ELSE
              PERFORM VALIDATE-DATA
           END-IF

         WHEN EIBAID = DFHPF5
           MOVE 'EMPPGCHG' TO PROGRAM-NAME
           PERFORM BRANCH-TO-PROGRAM

         WHEN EIBAID = DFHPF6
           MOVE 'EMPPGDEL' TO PROGRAM-NAME
           PERFORM BRANCH-TO-PROGRAM

         WHEN OTHER
           MOVE LOW-VALUES TO EMPADDO
           MOVE -1 TO EMPNOL
           MOVE "INVALID KEY PRESSED" TO MESSAGEO
           PERFORM SEND-MAP-DATA

       END-EVALUATE.

       EXEC CICS
          RETURN TRANSID('EMAD')
          COMMAREA (WS-COMMAREA)
       END-EXEC.

   PROCESS-PARA.

       PERFORM RECEIVE-MAP.
       INITIALIZE DCLEMPLOYEE MESSAGEO

       MOVE EMPNOI    TO WS-EMPNO
       MOVE WS-EMPNO  TO EMP-ID

       MOVE DFHBMFSE TO EMPNOF
       MOVE DFHBMFSE TO LNAMEF
```

```
        MOVE DFHBMFSE TO FNAMEF
        MOVE DFHBMFSE TO SOCSECF
        MOVE DFHBMFSE TO YRSSVCF
        MOVE DFHBMFSE TO LSTPRMF

        MOVE -1 TO EMPNOL.
        MOVE "ENTER DATA FOR NEW EMPLOYEE, THEN PRESS PF4 TO ADD"
            TO MESSAGEO

        MOVE "EMPPGADD" TO WS-PGM-PASS
        PERFORM SEND-MAP-ALL.

    VALIDATE-DATA.

        PERFORM RECEIVE-MAP
        INITIALIZE DCLEMPLOYEE MESSAGEO

        MOVE DFHBMFSE TO EMPNOF
        MOVE DFHBMFSE TO LNAMEF
        MOVE DFHBMFSE TO FNAMEF
        MOVE DFHBMFSE TO SOCSECF
        MOVE DFHBMFSE TO YRSSVCF
        MOVE DFHBMFSE TO LSTPRMF

        EVALUATE TRUE

            WHEN EMPNOI EQUAL SPACES OR EMPNOL EQUAL ZERO
                MOVE "EMPLOYEE NUMBER IS REQUIRED" TO MESSAGEO
                MOVE -1 TO EMPNOL

            WHEN EMPNOI IS NOT NUMERIC
                MOVE "EMPLOYEE NUMBER MUST BE NUMERIC" TO MESSAGEO
                MOVE -1 TO EMPNOL

            WHEN LNAMEI EQUAL SPACES OR LNAMEL EQUAL ZERO
                MOVE "EMPLOYEE LAST NAME IS REQUIRED" TO MESSAGEO
                MOVE -1 TO LNAMEL

            WHEN FNAMEI EQUAL SPACES OR FNAMEL EQUAL ZERO
                MOVE "EMPLOYEE FIRST NAME IS REQUIRED" TO MESSAGEO
                MOVE -1 TO FNAMEL

            WHEN SOCSECI EQUAL SPACES OR SOCSECL EQUAL ZERO
                MOVE "SOCIAL SECURITY NUMBER IS REQUIRED" TO MESSAGEO
                MOVE -1 TO SOCSECL

            WHEN SOCSECI IS NOT NUMERIC
                MOVE "SOCIAL SECURITY MUST BE NUMERIC" TO MESSAGEO
                MOVE -1 TO SOCSECL

            WHEN YRSSVCI EQUAL SPACES OR YRSSVCL EQUAL ZERO
                MOVE "YEARS OF SERVICE IS REQUIRED" TO MESSAGEO
                MOVE -1 TO YRSSVCL

            WHEN YRSSVCI IS NOT NUMERIC
                MOVE "YEARS OF SERVICE MUST BE NUMERIC" TO MESSAGEO
                MOVE -1 TO YRSSVCL
```

```
          WHEN LSTPRMI EQUAL SPACES OR LSTPRML EQUAL ZERO
              MOVE "LAST PROMOTION DATE IS REQUIRED" TO MESSAGEO
              MOVE -1 TO LSTPRML

          WHEN OTHER
              PERFORM ADD-RECORD
              MOVE -1 TO EMPNOL

      END-EVALUATE.

      PERFORM SEND-MAP-DATA.


  ADD-RECORD.

* MAP INPUT FIELDS TO VSAM RECORD STRUCTURE

      MOVE EMPNOI            TO WS-EMPNO
      MOVE WS-EMPNO          TO EMP-ID
      MOVE LNAMEI            TO EMP-LAST-NAME
      MOVE FNAMEI            TO EMP-FIRST-NAME
      MOVE SOCSECI           TO EMP-SSN
      MOVE YRSSVCI           TO WS-EMP-SRV-YRS
      MOVE WS-EMP-SRV-YRS    TO EMP-SERVICE-YEARS
      MOVE LSTPRMI           TO EMP-PROMOTION-DATE

* INSERT THE RECORD

      EXEC CICS
         WRITE
         FILE('EMPLOYEE')
         FROM(DCLEMPLOYEE)
         RIDFLD(EMP-ID)
         RESP(RESPONSE-CODE)
         RESP2(RESPONSE-CODE2)
      END-EXEC.

      EVALUATE RESPONSE-CODE

         WHEN DFHRESP(NORMAL)
             MOVE "EMPLOYEE ADDED SUCCESSFULLY" TO MESSAGEO
             MOVE -1 TO EMPNOL
             MOVE WS-EMPNO TO WS-EMP-PASS

         WHEN DFHRESP(DUPREC)
             MOVE "ERROR - RECORD ALREADY EXISTS" TO MESSAGEO
             MOVE -1 TO EMPNOL
             MOVE WS-EMPNO TO WS-EMP-PASS

         WHEN OTHER
             MOVE RESPONSE-CODE  TO RESPONSE-DISPLAY
             MOVE RESPONSE-CODE2 TO RESPONSE-DISPLA2
             STRING "UNKNOWN ERROR CODE: " DELIMITED BY SIZE
                 RESPONSE-DISPLAY DELIMITED BY SIZE
                 " "                 DELIMITED BY SIZE
                 RESPONSE-DISPLA2 DELIMITED BY SIZE
               INTO MESSAGEO
```

```
        END-EVALUATE.

BRANCH-TO-PROGRAM.

    EXEC CICS
        XCTL PROGRAM(PROGRAM-NAME)
        COMMAREA (WS-COMMAREA)
        LENGTH(10)
    END-EXEC

    MOVE 'PROGRAM NOT AVAILABLE' TO MESSAGEO.

SEND-MAP.
    EXEC CICS SEND
        MAP     ('EMPADD')
        MAPSET ('EMPMADD')
        FROM    (EMPADDO)
        CURSOR
        ERASE
    END-EXEC.

SEND-MAP-DATA.
    EXEC CICS SEND
        MAP     ('EMPADD')
        MAPSET ('EMPMADD')
        FROM    (EMPADDO)
        CURSOR
        DATAONLY
    END-EXEC.

SEND-MAP-ALL.
    EXEC CICS SEND
        MAP     ('EMPADD')
        MAPSET ('EMPMADD')
        FROM    (EMPADDO)
        CURSOR
    END-EXEC.

RECEIVE-MAP.
    EXEC CICS RECEIVE
        MAP     ('EMPADD')
        MAPSET ('EMPMADD')
        INTO    (EMPADDI)
    END-EXEC.
```

160

Ok, let's test the program. We'll add a record.

```
EMPMADD                      EMPLOYEE ADD                        EMAD

                  ENTER EMPLOYEE INFO, THEN PRESS PF4

        EMPLOYEE ID    9461

        EMP LAST NAME  berry

        EMP FIRST NAME julie

        EMP SOCIAL SEC 947294888

        EMP YEARS SRVC 34

        EMP LAST PROM  2018-01-01




    F2 INQ    F3 EXIT    F4 ADD    F5 CHG    F6 DEL
```

Press **ENTER** and this is the result, a successful add.

```
EMPMADD                      EMPLOYEE ADD                        EMAD

                  ENTER EMPLOYEE INFO, THEN PRESS PF4

        EMPLOYEE ID    9461

        EMP LAST NAME  BERRY

        EMP FIRST NAME JULIE

        EMP SOCIAL SEC 947294888

        EMP YEARS SRVC 34

        EMP LAST PROM  2018-01-01




EMPLOYEE ADDED SUCCESSFULLY
    F2 INQ    F3 EXIT    F4 ADD    F5 CHG    F6 DEL
```

Now we'll try adding the same record to test the duplicate record logic.

```
EMPMADD                    EMPLOYEE ADD                            EMAD


                    ENTER EMPLOYEE INFO, THEN PRESS PF4

         EMPLOYEE ID   9461

         EMP LAST NAME  BERRY

         EMP FIRST NAME JULIE

         EMP SOCIAL SEC 947294888

         EMP YEARS SRVC 34

         EMP LAST PROM  2018-01-01




    ERROR - RECORD ALREADY EXISTS
    F2 INQ   F3 EXIT   F4 ADD   F5 CHG   F6 DEL
```

Great, now go ahead and do your full unit test, then we'll do the change program.

EMPPGCHG

At this point we have everything we need to modify the change program. Simply copy the appropriate code from the inquiry and add programs (but remember to specify the UPDATE option on the read command – this is necessary to lock the record until you've finished updating it).

Go ahead and do these changes, and then we'll compare code.

……

Ok, back again. Did you run into any difficulties? I bet not, but just for comparison, here's my code.

```
IDENTIFICATION DIVISION.
PROGRAM-ID. EMPPGCHG.
****************************************************
*   COBOL/CICS/DB2 PROGRAM TO CHANGE AN EMPLOYEE *
*                                                *
*   AUTHOR       : ROBERT WINGATE                *
*   DATE-WRITTEN : 2018-07-23                    *
****************************************************
ENVIRONMENT DIVISION.
DATA DIVISION.
WORKING-STORAGE SECTION.
01 WS-EMPNO          PIC 9(4).
01 WS-EMP-SRV-YRS    PIC 9(02).
01 WS-SQLCODE        PIC 9(08).
01 WS-COMMAREA.
   05 WS-EMP-PASS    PIC 9(04).
   05 WS-PGM-PASS    PIC X(08).
   05 FILLER         PIC X(08).

01 RESPONSE-CODE     PIC S9(08) COMP.
01 RESPONSE-DISPLAY  PIC S9(08) USAGE DISPLAY.
01 RESPONSE-CODE2    PIC S9(08) VALUE 0.
01 RESPONSE-DISPLA2  PIC  9(08) VALUE 0.

01 PROGRAM-NAME      PIC X(08) VALUE SPACES.
01 SW-PASSED-DATA-SWITCH   PIC X(1) VALUE 'N'.
   88  SW-PASSED-DATA              VALUE 'Y'.
   88  SW-NO-PASSED-DATA           VALUE 'N'.

   COPY EMPMCHG.
   COPY DFHAID.
   COPY DFHBMSCA.

01 DCLEMPLOYEE.
   05 EMP-ID             PIC 9(04).
   05 EMP-LAST-NAME      PIC X(30).
   05 EMP-FIRST-NAME     PIC X(20).
   05 EMP-SERVICE-YEARS  PIC 9(02).
   05 EMP-PROMOTION-DATE PIC X(10).
   05 EMP-SSN            PIC X(09).
```

```
          05 FILLER               PIC X(05).

    LINKAGE SECTION.
    01 DFHCOMMAREA          PIC X(20).

    PROCEDURE DIVISION.

        SET SW-NO-PASSED-DATA TO TRUE

        IF EIBCALEN > ZERO
          MOVE DFHCOMMAREA  TO WS-COMMAREA
        END-IF.

        EVALUATE TRUE

          WHEN EIBCALEN = ZERO
            MOVE LOW-VALUES   TO  EMPCHGO
            MOVE -1 TO EMPINL
            PERFORM SEND-MAP

          WHEN EIBAID = DFHCLEAR
            MOVE LOW-VALUES   TO  EMPCHGO
            MOVE -1 TO EMPINL
            PERFORM SEND-MAP

          WHEN EIBAID = DFHPA1 OR DFHPA2 OR DFHPA3
            CONTINUE

          WHEN EIBAID = DFHPF2
            MOVE 'EMPPGINQ' TO PROGRAM-NAME
            PERFORM BRANCH-TO-PROGRAM

            EXEC CICS
              RETURN
            END-EXEC

          WHEN EIBAID = DFHPF3
            MOVE LOW-VALUES TO  EMPCHGO
            MOVE -1 TO EMPINL
            MOVE "BYE, PRESS CLEAR KEY TO ENTER A TRANSACTION ID"
                TO MESSAGEO
            PERFORM SEND-MAP-DATA

            EXEC CICS
              RETURN
            END-EXEC

          WHEN EIBAID = DFHPF4
            MOVE 'EMPPGADD' TO PROGRAM-NAME
            PERFORM BRANCH-TO-PROGRAM

            EXEC CICS
              RETURN
            END-EXEC

          WHEN EIBAID = DFHPF5
    *         PERFORM THE EDITS AND VALIDATIONS
    *         IF NO ERRORS THEN MODIFY THE RECORD
```

164

```
            IF WS-PGM-PASS NOT EQUAL "EMPPGCHG"
               SET SW-PASSED-DATA TO TRUE
               PERFORM PROCESS-PARA
            ELSE
               PERFORM VALIDATE-DATA
            END-IF

          WHEN EIBAID = DFHPF6
            MOVE 'EMPPGDEL' TO PROGRAM-NAME
            PERFORM BRANCH-TO-PROGRAM

            EXEC CICS
              RETURN
            END-EXEC

          WHEN EIBAID = DFHENTER
            PERFORM PROCESS-PARA

          WHEN OTHER
            MOVE LOW-VALUES TO EMPCHGO
            MOVE -1 TO EMPINL
            MOVE "INVALID KEY PRESSED" TO MESSAGEO
            PERFORM SEND-MAP-DATA

       END-EVALUATE.

       EXEC CICS
          RETURN TRANSID('EMCH')
          COMMAREA (WS-COMMAREA)
          LENGTH(20)
       END-EXEC.

   PROCESS-PARA.

       PERFORM RECEIVE-MAP.
       INITIALIZE DCLEMPLOYEE MESSAGEO

       IF SW-PASSED-DATA
          MOVE WS-EMP-PASS TO EMP-ID
       ELSE
          MOVE EMPINI    TO WS-EMPNO
          MOVE WS-EMPNO  TO EMP-ID
       END-IF

       MOVE DFHBMFSE TO EMPINF
       MOVE DFHBMFSE TO EMPNOF
       MOVE DFHBMFSE TO LNAMEF
       MOVE DFHBMFSE TO FNAMEF
       MOVE DFHBMFSE TO SOCSECF
       MOVE DFHBMFSE TO YRSSVCF
       MOVE DFHBMFSE TO LSTPRMF

       EXEC CICS
          READ
          FILE('EMPLOYEE')
          INTO(DCLEMPLOYEE)
          UPDATE
```

```
                RIDFLD(EMP-ID)
                EQUAL
                RESP(RESPONSE-CODE)
        END-EXEC.

        EVALUATE RESPONSE-CODE
            WHEN DFHRESP(NORMAL)
                MOVE EMP-ID                 TO  WS-EMPNO WS-EMP-PASS
                MOVE WS-EMPNO               TO  EMPNOO
                MOVE EMP-LAST-NAME          TO  LNAMEO
                MOVE EMP-FIRST-NAME         TO  FNAMEO
                MOVE EMP-SSN                TO  SOCSECO
                MOVE EMP-SERVICE-YEARS      TO WS-EMP-SRV-YRS
                MOVE WS-EMP-SRV-YRS         TO  YRSSVCO
                MOVE EMP-PROMOTION-DATE     TO LSTPRMO


                MOVE "MAKE CHANGES AND THEN PRESS PF5" TO MESSAGEO

            WHEN DFHRESP(NOTFND)
                STRING "EMPLOYEE ID " DELIMITED BY SIZE
                WS-EMPNO DELIMITED BY SPACE
                " NOT FOUND" DELIMITED BY SIZE INTO MESSAGEO
                MOVE SPACES        TO EMPNOO
                MOVE SPACES        TO LNAMEO
                MOVE SPACES        TO FNAMEO
                MOVE SPACES        TO SOCSECO
                MOVE SPACES        TO YRSSVCO
                MOVE SPACES        TO LSTPRMO

            WHEN OTHER
                MOVE RESPONSE-CODE TO RESPONSE-DISPLAY
                STRING "UNKNOWN ERROR CODE: " DELIMITED BY SIZE
                    RESPONSE-DISPLAY DELIMITED BY SIZE
                  INTO MESSAGEO


        END-EVALUATE.

        MOVE -1 TO EMPINL.
        MOVE "EMPPGCHG" TO WS-PGM-PASS
        IF SW-PASSED-DATA
            PERFORM SEND-MAP
        ELSE
            PERFORM SEND-MAP-DATA
        END-IF.

    VALIDATE-DATA.

        PERFORM RECEIVE-MAP
        INITIALIZE DCLEMPLOYEE MESSAGEO

        EVALUATE TRUE
            WHEN EMPNOI EQUAL SPACES OR EMPNOL EQUAL ZERO
                MOVE "EMPLOYEE NUMBER IS REQUIRED" TO MESSAGEO
                MOVE -1 TO EMPNOL

            WHEN EMPNOI IS NOT NUMERIC
                MOVE "EMPLOYEE NUMBER MUST BE NUMERIC" TO MESSAGEO
                MOVE -1 TO EMPNOL
```

166

```
            WHEN LNAMEI EQUAL SPACES OR LNAMEL EQUAL ZERO
                MOVE "EMPLOYEE LAST NAME IS REQUIRED" TO MESSAGEO
                MOVE -1 TO LNAMEL

            WHEN FNAMEI EQUAL SPACES OR FNAMEL EQUAL ZERO
                MOVE "EMPLOYEE FIRST NAME IS REQUIRED" TO MESSAGEO
                MOVE -1 TO FNAMEL

            WHEN SOCSECI EQUAL SPACES OR SOCSECL EQUAL ZERO
                MOVE "SOCIAL SECURITY NUMBER IS REQUIRED" TO MESSAGEO
                MOVE -1 TO SOCSECL

            WHEN SOCSECI IS NOT NUMERIC
                MOVE "SOCIAL SECURITY MUST BE NUMERIC" TO MESSAGEO
                MOVE -1 TO SOCSECL

            WHEN YRSSVCI EQUAL SPACES OR YRSSVCL EQUAL ZERO
                MOVE "YEARS OF SERVICE IS REQUIRED" TO MESSAGEO
                MOVE -1 TO YRSSVCL

            WHEN YRSSVCI IS NOT NUMERIC
                MOVE "YEARS OF SERVICE MUST BE NUMERIC" TO MESSAGEO
                MOVE -1 TO YRSSVCL

            WHEN LSTPRMI EQUAL SPACES OR LSTPRML EQUAL ZERO
                MOVE "LAST PROMOTION DATE IS REQUIRED" TO MESSAGEO
                MOVE -1 TO LSTPRML

            WHEN OTHER
                PERFORM CHANGE-RECORD
                MOVE -1 TO EMPINL

        END-EVALUATE.

        MOVE DFHBMFSE TO EMPINF
        MOVE DFHBMFSE TO EMPNOF
        MOVE DFHBMFSE TO LNAMEF
        MOVE DFHBMFSE TO FNAMEF
        MOVE DFHBMFSE TO SOCSECF
        MOVE DFHBMFSE TO YRSSVCF
        MOVE DFHBMFSE TO LSTPRMF

        MOVE "EMPPGCHG" TO WS-PGM-PASS
        PERFORM SEND-MAP-DATA.

    CHANGE-RECORD.

    * FIRST REGET THE RECORD FOR UPDATE

        MOVE EMPINI    TO WS-EMPNO
        MOVE WS-EMPNO  TO EMP-ID

        EXEC CICS
            READ
            FILE('EMPLOYEE')
            INTO(DCLEMPLOYEE)
            UPDATE
```

```
            RIDFLD(EMP-ID)
            EQUAL
            RESP(RESPONSE-CODE)
      END-EXEC.

      IF RESPONSE-CODE EQUAL DFHRESP(NORMAL)

*         DO THE UPDATE

          MOVE EMPNOI              TO WS-EMPNO
          MOVE WS-EMPNO            TO EMP-ID
          MOVE LNAMEI              TO EMP-LAST-NAME
          MOVE FNAMEI              TO EMP-FIRST-NAME
          MOVE SOCSECI             TO EMP-SSN
          MOVE YRSSVCI             TO WS-EMP-SRV-YRS
          MOVE WS-EMP-SRV-YRS      TO EMP-SERVICE-YEARS
          MOVE LSTPRMI             TO EMP-PROMOTION-DATE

      EXEC CICS
          REWRITE
          FILE('EMPLOYEE')
          FROM(DCLEMPLOYEE)
          RESP(RESPONSE-CODE)
          RESP2(RESPONSE-CODE2)
      END-EXEC

      EVALUATE RESPONSE-CODE

          WHEN DFHRESP(NORMAL)
             MOVE "EMPLOYEE MODIFED SUCCESSFULLY" TO MESSAGEO
             MOVE -1 TO EMPNOL
             MOVE WS-EMPNO TO WS-EMP-PASS

          WHEN DFHRESP(NOTFND)
             MOVE -1 TO EMPNOL
             MOVE WS-EMPNO TO WS-EMP-PASS
             STRING "EMPLOYEE ID " DELIMITED BY SIZE
             WS-EMPNO DELIMITED BY SPACE
             " NOT FOUND" DELIMITED BY SIZE INTO MESSAGEO
             MOVE SPACES         TO EMPNOO
             MOVE SPACES         TO LNAMEO
             MOVE SPACES         TO FNAMEO
             MOVE SPACES         TO SOCSECO
             MOVE SPACES         TO YRSSVCO
             MOVE SPACES         TO LSTPRMO
             MOVE -1 TO EMPNOL

          WHEN OTHER
             MOVE RESPONSE-CODE  TO RESPONSE-DISPLAY
             MOVE RESPONSE-CODE2 TO RESPONSE-DISPLA2
             STRING "UNKNOWN ERROR CODE: " DELIMITED BY SIZE
                 RESPONSE-DISPLAY DELIMITED BY SIZE
                 " "                DELIMITED BY SIZE
                 RESPONSE-DISPLA2 DELIMITED BY SIZE
               INTO MESSAGEO

      END-EVALUATE
```

```
              ELSE

                  MOVE RESPONSE-CODE TO RESPONSE-DISPLAY
                  STRING "UNKNOWN ERROR CODE: " DELIMITED BY SIZE
                       RESPONSE-DISPLAY DELIMITED BY SIZE
                    INTO MESSAGEO

              END-IF.

      BRANCH-TO-PROGRAM.
          EXEC CICS
              XCTL PROGRAM(PROGRAM-NAME)
              COMMAREA (WS-COMMAREA)
              LENGTH(20)
          END-EXEC

          MOVE 'PROGRAM NOT AVAILABLE' TO MESSAGEO.

      SEND-MAP.
          EXEC CICS SEND
              MAP    ('EMPCHG')
              MAPSET ('EMPMCHG')
              FROM   (EMPCHGO)
              CURSOR
              ERASE
          END-EXEC.

      SEND-MAP-DATA.
          EXEC CICS SEND
              MAP    ('EMPCHG')
              MAPSET ('EMPMCHG')
              FROM   (EMPCHGO)
              DATAONLY
              CURSOR
          END-EXEC.

      SEND-MAP-ALL.
          EXEC CICS SEND
              MAP    ('EMPCHG')
              MAPSET ('EMPMCHG')
              FROM   (EMPCHGO)
              CURSOR
          END-EXEC.

      RECEIVE-MAP.
           EXEC CICS RECEIVE
              MAP    ('EMPCHG')
              MAPSET ('EMPMCHG')
              INTO   (EMPCHGI)
           END-EXEC.
```

Now let's change a record and see if our changes took. Let's pick employee 3217, and we'll change the years of service to 7 and the last promotion date to 2018-01-01. First, bring up the change screen for employee 3217:

```
EMPMCHG                    EMPLOYEE CHANGE                         EMCH


        EMPLOYEE ->    3217       ENTER EMPLOYEE ID, THEN PRESS ENTER

        EMPLOYEE ID    3217

        EMP LAST NAME  JOHNSON

        EMP FIRST NAME EDWARD

        EMP SOCIAL SEC 493082938

        EMP YEARS SRVC 04

        EMP LAST PROM  2017-01-01




MAKE CHANGES AND THEN PRESS PF5
F2 INQ   F3 EXIT   F4 ADD   F5 CHG   F6 DEL
```

Now make the changes and press PF5 to do the update.

```
EMPMCHG                    EMPLOYEE CHANGE                         EMCH


        EMPLOYEE ->    3217       ENTER EMPLOYEE ID, THEN PRESS ENTER

        EMPLOYEE ID    3217

        EMP LAST NAME  JOHNSON

        EMP FIRST NAME EDWARD

        EMP SOCIAL SEC 493082938

        EMP YEARS SRVC 07

        EMP LAST PROM  2018-01-01




EMPLOYEE MODIFED SUCCESSFULLY
F2 INQ   F3 EXIT   F4 ADD   F5 CHG   F6 DEL
```

Also, let's check to make sure the error logic still works if we enter a non-existent employee id.

```
EMPMCHG                    EMPLOYEE CHANGE                         EMCH


      EMPLOYEE ->    4321      ENTER EMPLOYEE ID, THEN PRESS ENTER

      EMPLOYEE ID

      EMP LAST NAME

      EMP FIRST NAME

      EMP SOCIAL SEC

      EMP YEARS SRVC

      EMP LAST PROM




  EMPLOYEE ID 4321 NOT FOUND
  F2 INQ    F3 EXIT    F4 ADD    F5 CHG    F6 DEL
```

Excellent, now go ahead and do the full unit test. Then we'll finish with the delete program.

EMPPGDEL

The delete program is easy to change. We just need to change the initial inquiry for display, and then code the delete after the user presses PF6. At this point you have everything you need to do the coding. Go ahead and give it a try. Then take a good coffee/tea/soda break and come back and compare code.

......

Ok, time to compare code. Here is mine:

```
      IDENTIFICATION DIVISION.
      PROGRAM-ID. EMPPGDEL.
     ***********************************************
     *  COBOL/CICS/DB2 PROGRAM TO DELETE AN EMPLOYEE *
     *                                               *
     *  AUTHOR       : ROBERT WINGATE                *
     *  DATE-WRITTEN : 2018-07-25                    *
     ***********************************************
      ENVIRONMENT DIVISION.
      DATA DIVISION.
```

```
WORKING-STORAGE SECTION.
01 WS-EMPNO          PIC 9(4).
01 WS-EMP-SRV-YRS    PIC 9(2).
01 WS-COMMAREA.
   05 WS-EMP-PASS    PIC 9(04).
   05 WS-PGM-PASS    PIC X(08).
   05 FILLER         PIC X(08).

01 RESPONSE-CODE     PIC S9(08) VALUE 0.
01 RESPONSE-DISPLAY  PIC  9(08) VALUE 0.
01 RESPONSE-CODE2    PIC S9(08) VALUE 0.
01 RESPONSE-DISPLA2  PIC  9(08) VALUE 0.

01 PROGRAM-NAME      PIC X(08) VALUE SPACES.
01 SW-PASSED-DATA-SWITCH   PIC X(1) VALUE 'N'.
   88  SW-PASSED-DATA                VALUE 'Y'.
   88  SW-NO-PASSED-DATA             VALUE 'N'.

   COPY EMPMDEL.
   COPY DFHAID.
   COPY DFHBMSCA.

01 DCLEMPLOYEE.
   05 EMP-ID              PIC 9(04).
   05 EMP-LAST-NAME       PIC X(30).
   05 EMP-FIRST-NAME      PIC X(20).
   05 EMP-SERVICE-YEARS   PIC 9(02).
   05 EMP-PROMOTION-DATE  PIC X(10).
   05 EMP-SSN             PIC X(09).
   05 FILLER              PIC X(05).

LINKAGE SECTION.

01 DFHCOMMAREA           PIC X(20).

PROCEDURE DIVISION.

   IF EIBCALEN > ZERO
     MOVE DFHCOMMAREA  TO WS-COMMAREA
   END-IF.

   EVALUATE TRUE

     WHEN EIBCALEN = ZERO
       MOVE LOW-VALUES   TO  EMPDELO
       PERFORM SEND-MAP

     WHEN EIBAID = DFHCLEAR
       MOVE LOW-VALUES    TO  EMPDELO
       PERFORM SEND-MAP

     WHEN EIBAID = DFHPA1 OR DFHPA2 OR DFHPA3
       CONTINUE

     WHEN EIBAID = DFHPF2
       MOVE 'EMPPGINQ' TO PROGRAM-NAME
       PERFORM BRANCH-TO-PROGRAM
```

```
          WHEN EIBAID = DFHPF3
            MOVE LOW-VALUES TO  EMPDELO
            MOVE "BYE, PRESS CLEAR KEY TO ENTER A TRANSACTION ID"
                 TO MESSAGEO
            PERFORM SEND-MAP-DATA

            EXEC CICS
              RETURN
            END-EXEC

          WHEN EIBAID = DFHPF4
            MOVE 'EMPPGADD' TO PROGRAM-NAME
            PERFORM BRANCH-TO-PROGRAM

          WHEN EIBAID = DFHPF5
            MOVE 'EMPPGCHG' TO PROGRAM-NAME
            PERFORM BRANCH-TO-PROGRAM

          WHEN EIBAID = DFHPF6

            IF WS-PGM-PASS NOT EQUAL "EMPPGDEL"
               SET SW-PASSED-DATA TO TRUE
               PERFORM PROCESS-PARA
            ELSE
               SET SW-NO-PASSED-DATA TO TRUE
               PERFORM DELETE-RECORD
            END-IF

          WHEN EIBAID = DFHENTER
            PERFORM PROCESS-PARA

          WHEN OTHER
            MOVE LOW-VALUES TO EMPDELO
            MOVE "INVALID KEY PRESSED" TO MESSAGEO
            PERFORM SEND-MAP-DATA

      END-EVALUATE.

      EXEC CICS
         RETURN TRANSID('EMDE')
         COMMAREA (WS-COMMAREA)
      END-EXEC.

  PROCESS-PARA.

      PERFORM RECEIVE-MAP.
      INITIALIZE DCLEMPLOYEE MESSAGEO

      IF SW-PASSED-DATA
         MOVE WS-EMP-PASS TO WS-EMPNO
      ELSE
         MOVE EMPINI      TO WS-EMPNO
      END-IF

      MOVE WS-EMPNO  TO EMP-ID

      EXEC CICS
         READ
```

```
           FILE('EMPLOYEE')
           INTO(DCLEMPLOYEE)
           RIDFLD(EMP-ID)
           EQUAL
           RESP(RESPONSE-CODE)
       END-EXEC.

       EVALUATE RESPONSE-CODE

           WHEN DFHRESP(NORMAL)
               MOVE EMP-ID                TO   WS-EMPNO WS-EMP-PASS
               MOVE WS-EMPNO              TO   EMPNOO
               MOVE EMP-LAST-NAME         TO LNAMEO
               MOVE EMP-FIRST-NAME        TO FNAMEO
               MOVE EMP-SSN               TO SOCSECO
               MOVE EMP-SERVICE-YEARS     TO WS-EMP-SRV-YRS
               MOVE WS-EMP-SRV-YRS        TO YRSSVCO
               MOVE EMP-PROMOTION-DATE    TO LSTPRMO
               MOVE "PRESS PF6 TO DELETE THIS RECORD" TO MESSAGEO

           WHEN DFHRESP(NOTFND)
               STRING "EMPLOYEE ID " DELIMITED BY SIZE
               WS-EMPNO DELIMITED BY SPACE
               " NOT FOUND" DELIMITED BY SIZE INTO MESSAGEO
               MOVE SPACES        TO EMPNOO
               MOVE SPACES        TO LNAMEO
               MOVE SPACES        TO FNAMEO
               MOVE SPACES        TO SOCSECO
               MOVE SPACES        TO YRSSVCO
               MOVE SPACES        TO LSTPRMO

           WHEN OTHER
               MOVE RESPONSE-CODE TO RESPONSE-DISPLAY
               STRING "UNKNOWN ERROR CODE: " DELIMITED BY SIZE
                   RESPONSE-DISPLAY DELIMITED BY SIZE
                 INTO MESSAGEO

       END-EVALUATE.

       MOVE DFHBMFSE TO EMPINF
       MOVE -1 TO EMPINL.

       MOVE EMP-ID     TO WS-EMP-PASS
       MOVE "EMPPGDEL" TO WS-PGM-PASS

       IF SW-PASSED-DATA
          PERFORM SEND-MAP
       ELSE
          PERFORM SEND-MAP-DATA
       END-IF.

   DELETE-RECORD.

       PERFORM RECEIVE-MAP
       INITIALIZE DCLEMPLOYEE MESSAGEO

*  MAP INPUT FIELDS TO DB2 RECORD
```

```
        MOVE EMPINI            TO WS-EMPNO
        MOVE WS-EMPNO          TO EMP-ID

*   DELETE THE RECORD

        EXEC CICS
           DELETE
           FILE('EMPLOYEE')
           RIDFLD(EMP-ID)
           RESP(RESPONSE-CODE)
           RESP2(RESPONSE-CODE2)
        END-EXEC

        EVALUATE RESPONSE-CODE

           WHEN DFHRESP(NORMAL)
               MOVE "EMPLOYEE DELETED SUCCESSFULLY" TO MESSAGEO
               MOVE -1 TO EMPNOL
               MOVE WS-EMPNO TO WS-EMP-PASS

           WHEN DFHRESP(NOTFND)
               MOVE -1 TO EMPNOL
               MOVE WS-EMPNO TO WS-EMP-PASS
               STRING "EMPLOYEE ID " DELIMITED BY SIZE
               WS-EMPNO DELIMITED BY SPACE
               " NOT FOUND" DELIMITED BY SIZE INTO MESSAGEO
               MOVE SPACES         TO EMPNOO
               MOVE SPACES         TO LNAMEO
               MOVE SPACES         TO FNAMEO
               MOVE SPACES         TO SOCSECO
               MOVE SPACES         TO YRSSVCO
               MOVE SPACES         TO LSTPRMO
               MOVE -1 TO EMPNOL

           WHEN OTHER
               MOVE RESPONSE-CODE  TO RESPONSE-DISPLAY
               MOVE RESPONSE-CODE2 TO RESPONSE-DISPLA2
               STRING "UNKNOWN ERROR CODE: " DELIMITED BY SIZE
                   RESPONSE-DISPLAY DELIMITED BY SIZE
                   " "              DELIMITED BY SIZE
                   RESPONSE-DISPLA2 DELIMITED BY SIZE
                 INTO MESSAGEO

        END-EVALUATE

        MOVE DFHBMFSE TO EMPINF
        MOVE -1 TO EMPINL.
        MOVE "EMPPGDEL" TO WS-PGM-PASS
        PERFORM SEND-MAP-DATA.

   BRANCH-TO-PROGRAM.
        EXEC CICS
           XCTL PROGRAM(PROGRAM-NAME)
           COMMAREA (WS-COMMAREA)
           LENGTH(10)
        END-EXEC

        MOVE 'PROGRAM NOT AVAILABLE' TO MESSAGEO.
```

175

```
SEND-MAP.
    EXEC CICS SEND
        MAP     ('EMPDEL')
        MAPSET ('EMPMDEL')
        FROM    (EMPDELO)
        ERASE
    END-EXEC.

SEND-MAP-DATA.
    EXEC CICS SEND
        MAP     ('EMPDEL')
        MAPSET ('EMPMDEL')
        FROM    (EMPDELO)
        DATAONLY
    END-EXEC.

SEND-MAP-ALL.
    EXEC CICS SEND
        MAP     ('EMPDEL')
        MAPSET ('EMPMDEL')
        FROM    (EMPDELO)
    END-EXEC.

RECEIVE-MAP.
    EXEC CICS RECEIVE
        MAP     ('EMPDEL')
        MAPSET ('EMPMDEL')
        INTO    (EMPDELI)
    END-EXEC.
```

Compile and link the program, remember to refresh it in CICS with CEMT. Now let's test it. Bring up employee number 9999 on the delete screen.

```
EMPMDEL                     EMPLOYEE DELETE                         EMDE


        EMPLOYEE ->    9999       ENTER EMPLOYEE ID, THEN PRESS ENTER

        EMPLOYEE ID    9999

        EMP LAST NAME  WINGATE

        EMP FIRST NAME ROBERT

        EMP SOCIAL SEC 999999999

        EMP YEARS SRVC 22

        EMP LAST PROM  2018-01-01




PRESS PF6 TO DELETE THIS RECORD
F2 INQ   F3 EXIT   F4 ADD   F5 CHG   F6 DEL
```

Press PF6 to get the result:

```
EMPMDEL                   EMPLOYEE DELETE                        EMDE

       EMPLOYEE ->    9999       ENTER EMPLOYEE ID, THEN PRESS ENTER

       EMPLOYEE ID    9999

       EMP LAST NAME  WINGATE

       EMP FIRST NAME ROBERT

       EMP SOCIAL SEC 999999999

       EMP YEARS SRVC 22

       EMP LAST PROM  2018-01-01




 EMPLOYEE DELETED SUCCESSFULLY
 F2 INQ    F3 EXIT    F4 ADD    F5 CHG    F6 DEL
```

Finally, go ahead and complete the full unit test for EMPPGDEL. Then we'll complete our project with some integration testing.

Integration Testing

EMPPGMU

Ok, let's repeat our integration test, beginning with the menu program **EMPPGMNU**. We'll simply verify that we can still navigate to all the screens and that they display properly. We'll choose the menu options for inquiry, add, change and delete. First inquiry.

```
EMPMMNU                       EMPLOYEE SUPPORT MENU                    EMNU

          ENTER THE NUMBER OF YOUR SELECTION,   THEN PRESS ENTER.


                     1   1. EMPLOYEE INQUIRY

                         2. EMPLOYEE ADD

                         3. EMPLOYEE CHANGE

                         4. EMPLOYEE DELETE





F3 EXIT


EMPMINQ                       EMPLOYEE INQUIRY                         EMIN

        EMPLOYEE ->               ENTER EMPLOYEE ID, THEN PRESS ENTER

        EMPLOYEE ID    XXXX

        EMP LAST NAME  XXXX

        EMP FIRST NAME XXXX

        EMP SOCIAL SEC XXXXXXXXX

        EMP YEARS SRVC 00

        EMP LAST PROM  YYYY-MM-DD




F2 INQ   F3 EXIT   F4 ADD   F5 CHG   F6 DEL
```

Next choose the add option.

```
EMPMMNU                    EMPLOYEE SUPPORT MENU                        EMNU
            ENTER THE NUMBER OF YOUR SELECTION,   THEN PRESS ENTER.


                    2    1. EMPLOYEE INQUIRY

                         2. EMPLOYEE ADD

                         3. EMPLOYEE CHANGE

                         4. EMPLOYEE DELETE







F3 EXIT
```

```
EMPMADD                      EMPLOYEE ADD                               EMAD

                    ENTER EMPLOYEE INFO, THEN PRESS PF4

        EMPLOYEE ID

        EMP LAST NAME

        EMP FIRST NAME

        EMP SOCIAL SEC

        EMP YEARS SRVC

        EMP LAST PROM




F2 INQ   F3 EXIT   F4 ADD   F5 CHG   F6 DEL
```

Now return to the main menu and choose the change option.

```
EMPMMNU                    EMPLOYEE SUPPORT MENU                    EMNU

            ENTER THE NUMBER OF YOUR SELECTION,   THEN PRESS ENTER.


                         3   1. EMPLOYEE INQUIRY

                             2. EMPLOYEE ADD

                             3. EMPLOYEE CHANGE

                             4. EMPLOYEE DELETE








    F3 EXIT



EMPMCHG                      EMPLOYEE CHANGE                        EMCH


        EMPLOYEE ->     ____      ENTER EMPLOYEE ID, THEN PRESS ENTER

        EMPLOYEE ID    XXXX

        EMP LAST NAME  XXXX

        EMP FIRST NAME XXXX

        EMP SOCIAL SEC XXXXXXXX

        EMP YEARS SRVC 00

        EMP LAST PROM  YYYY-MM-DD




    F2 INQ   F3 EXIT   F4 ADD   F5 CHG   F6 DEL
```

Finally, return to the main menu and choose the delete option.

```
EMPMMNU                     EMPLOYEE SUPPORT MENU                    EMNU

          ENTER THE NUMBER OF YOUR SELECTION,  THEN PRESS ENTER.


                    4   1. EMPLOYEE INQUIRY

                        2. EMPLOYEE ADD

                        3. EMPLOYEE CHANGE

                        4. EMPLOYEE DELETE







   F3 EXIT
```

```
EMPMDEL                      EMPLOYEE DELETE                         EMDE


       EMPLOYEE ->              ENTER EMPLOYEE ID, THEN PRESS ENTER

       EMPLOYEE ID    XXXX

       EMP LAST NAME  XXXX

       EMP FIRST NAME XXXX

       EMP SOCIAL SEC XXXXXXXXX

       EMP YEARS SRVC 00

       EMP LAST PROM  YYYY-MM-DD




   F2 INQ   F3 EXIT   F4 ADD   F5 CHG   F6 DEL
```

All looks good. Now on to the data operations.

EMPPGINQ

Ok let's start out with the primary display and entry after choosing from the main menu. Let's use 7777 as the employee id.

```
EMPMINQ                    EMPLOYEE INQUIRY                         EMIN


        EMPLOYEE ->    7777      ENTER EMPLOYEE ID, THEN PRESS ENTER

        EMPLOYEE ID    XXXX

        EMP LAST NAME  XXXX

        EMP FIRST NAME XXXX

        EMP SOCIAL SEC XXXXXXXXX

        EMP YEARS SRVC 00

        EMP LAST PROM  YYYY-MM-DD




     F2 INQ   F3 EXIT   F4 ADD   F5 CHG   F6 DEL


EMPMINQ                    EMPLOYEE INQUIRY                         EMIN


        EMPLOYEE ->    7777      ENTER EMPLOYEE ID, THEN PRESS ENTER

        EMPLOYEE ID    7777

        EMP LAST NAME  JACKSON

        EMP FIRST NAME JOSEPH

        EMP SOCIAL SEC 382746236

        EMP YEARS SRVC 17

        EMP LAST PROM  2017-01-01




     F2 INQ   F3 EXIT   F4 ADD   F5 CHG   F6 DEL
```

This looks good. Now let's try passing the same employee number from another program, such as the change program. Bring up employee 7777 again, and then press PF2 to transfer to the inquiry program.

```
EMPMCHG                    EMPLOYEE CHANGE                              EMCH


        EMPLOYEE ->    7777      ENTER EMPLOYEE ID, THEN PRESS ENTER

        EMPLOYEE ID    7777

        EMP LAST NAME  JACKSON

        EMP FIRST NAME JOSEPH

        EMP SOCIAL SEC 382746236

        EMP YEARS SRVC 17

        EMP LAST PROM  2017-01-01




  MAKE CHANGES AND THEN PRESS PF5
  F2 INQ   F3 EXIT   F4 ADD   F5 CHG   F6 DEL



EMPMINQ                    EMPLOYEE INQUIRY                             EMIN


        EMPLOYEE ->    7777      ENTER EMPLOYEE ID, THEN PRESS ENTER

        EMPLOYEE ID    7777

        EMP LAST NAME  JACKSON

        EMP FIRST NAME JOSEPH

        EMP SOCIAL SEC 382746236

        EMP YEARS SRVC 17

        EMP LAST PROM  2017-01-01




  F2 INQ   F3 EXIT   F4 ADD   F5 CHG   F6 DEL
```

Finally we should test the transfers from the inquiry screen to the add, change and delete screens. Let's do that to make sure the program interfaces are working correctly. Start with inquiry to add.

```
EMPMINQ                    EMPLOYEE INQUIRY                        EMIN


        EMPLOYEE ->   7777       ENTER EMPLOYEE ID, THEN PRESS ENTER

        EMPLOYEE ID   7777

        EMP LAST NAME  JACKSON

        EMP FIRST NAME JOSEPH

        EMP SOCIAL SEC 382746236

        EMP YEARS SRVC 17

        EMP LAST PROM  2017-01-01




   F2 INQ   F3 EXIT   F4 ADD   F5 CHG   F6 DEL
EMPMADD                    EMPLOYEE ADD                           EMAD

                    ENTER EMPLOYEE INFO, THEN PRESS PF4

        EMPLOYEE ID

        EMP LAST NAME

        EMP FIRST NAME

        EMP SOCIAL SEC

        EMP YEARS SRVC

        EMP LAST PROM




   ENTER DATA FOR NEW EMPLOYEE, THEN PRESS PF4 TO ADD
   F2 INQ   F3 EXIT   F4 ADD   F5 CHG   F6 DEL
```

Also go ahead and add an employee to ensure all attribute setting, switches and variables are correct.

```
EMPMADD                  EMPLOYEE ADD                           EMAD

                 ENTER EMPLOYEE INFO, THEN PRESS PF4

        EMPLOYEE ID    1111

        EMP LAST NAME  stone

        EMP FIRST NAME steven

        EMP SOCIAL SEC 385610088

        EMP YEARS SRVC 12

        EMP LAST PROM  2016-01-01




   ENTER DATA FOR NEW EMPLOYEE, THEN PRESS PF4 TO ADD
   F2 INQ   F3 EXIT   F4 ADD   F5 CHG   F6 DEL
```

```
EMPMADD                  EMPLOYEE ADD                           EMAD

                 ENTER EMPLOYEE INFO, THEN PRESS PF4

        EMPLOYEE ID    1111

        EMP LAST NAME  STONE

        EMP FIRST NAME STEVEN

        EMP SOCIAL SEC 385610088

        EMP YEARS SRVC 12

        EMP LAST PROM  2016-01-01




   EMPLOYEE ADDED SUCCESSFULLY
   F2 INQ   F3 EXIT   F4 ADD   F5 CHG   F6 DEL
```

All looks well, so let's move on to the add screen.

EMPPGADD

We just added a record, so let's try transferring to the inquiry screen, the change screen and then the delete screen. Press PF2.

```
EMPMINQ                    EMPLOYEE INQUIRY                         EMIN


        EMPLOYEE ->              ENTER EMPLOYEE ID, THEN PRESS ENTER

        EMPLOYEE ID    1111

        EMP LAST NAME   STONE

        EMP FIRST NAME STEVEN

        EMP SOCIAL SEC 385610088

        EMP YEARS SRVC 12

        EMP LAST PROM  2016-01-01




     F2 INQ   F3 EXIT   F4 ADD   F5 CHG   F6 DEL

EMPMINQ                    EMPLOYEE INQUIRY                         EMIN


        EMPLOYEE ->              ENTER EMPLOYEE ID, THEN PRESS ENTER

        EMPLOYEE ID    1111

        EMP LAST NAME   STONE

        EMP FIRST NAME STEVEN

        EMP SOCIAL SEC 385610088

        EMP YEARS SRVC 12

        EMP LAST PROM  2016-01-01




     F2 INQ   F3 EXIT   F4 ADD   F5 CHG   F6 DEL
```

Next let's add two more records, and transfer to the change and delete screens, respectively.

```
EMPMADD                    EMPLOYEE ADD

                    ENTER EMPLOYEE INFO, THEN PRESS PF4

        EMPLOYEE ID   1212

        EMP LAST NAME  SAMPLE

        EMP FIRST NAME RECORD

        EMP SOCIAL SEC 373737373

        EMP YEARS SRVC 04

        EMP LAST PROM  2017-01-01




  EMPLOYEE ADDED SUCCESSFULLY
  F2 INQ   F3 EXIT   F4 ADD   F5 CHG   F6 DEL


EMPMCHG                  EMPLOYEE CHANGE                    EMCH

        EMPLOYEE ->          ENTER EMPLOYEE ID, THEN PRESS ENTER

        EMPLOYEE ID   1212

        EMP LAST NAME  SAMPLE

        EMP FIRST NAME RECORD

        EMP SOCIAL SEC 373737373

        EMP YEARS SRVC 04

        EMP LAST PROM  2017-01-01




  MAKE CHANGES AND THEN PRESS PF5
  F2 INQ   F3 EXIT   F4 ADD   F5 CHG   F6 DEL
```

```
EMPMADD                    EMPLOYEE ADD                              EMAD


                    ENTER EMPLOYEE INFO, THEN PRESS PF4

        EMPLOYEE ID    2424

        EMP LAST NAME  TWO

        EMP FIRST NAME SAMPLE

        EMP SOCIAL SEC 747474747

        EMP YEARS SRVC 13

        EMP LAST PROM  2016-01-01




    EMPLOYEE ADDED SUCCESSFULLY
    F2 INQ   F3 EXIT   F4 ADD   F5 CHG   F6 DEL



EMPMDEL                    EMPLOYEE DELETE                           EMDE


        EMPLOYEE ->              ENTER EMPLOYEE ID, THEN PRESS ENTER

        EMPLOYEE ID    2424

        EMP LAST NAME  TWO

        EMP FIRST NAME SAMPLE

        EMP SOCIAL SEC 747474747

        EMP YEARS SRVC 13

        EMP LAST PROM  2016-01-01




    PRESS PF6 TO DELETE THIS RECORD
    F2 INQ   F3 EXIT   F4 ADD   F5 CHG   F6 DEL
```

All looks good with transferring from the add screen. Let's move on to the change screen.

EMPPGCHG

Let's bring up the 2424 record that we just created on the change screen. Now transfer to the inquiry screen, the delete screen, and finally the add screen (the latter will not process any transferred data except the program name).

```
EMPMCHG                    EMPLOYEE CHANGE                        EMCH


        EMPLOYEE ->    2424      ENTER EMPLOYEE ID, THEN PRESS ENTER

        EMPLOYEE ID    2424

        EMP LAST NAME  TWO

        EMP FIRST NAME SAMPLE

        EMP SOCIAL SEC 747474747

        EMP YEARS SRVC 13

        EMP LAST PROM  2016-01-01




    MAKE CHANGES AND THEN PRESS PF5
    F2 INQ   F3 EXIT   F4 ADD   F5 CHG   F6 DEL



EMPMINQ                    EMPLOYEE INQUIRY                       EMIN


        EMPLOYEE ->    2424       ENTER EMPLOYEE ID, THEN PRESS ENTER

        EMPLOYEE ID    2424

        EMP LAST NAME  TWO

        EMP FIRST NAME SAMPLE

        EMP SOCIAL SEC 747474747

        EMP YEARS SRVC 13

        EMP LAST PROM  2016-01-01




    F2 INQ   F3 EXIT   F4 ADD   F5 CHG   F6 DEL
```

```
EMPMCHG                    EMPLOYEE CHANGE                         EMCH


        EMPLOYEE ->    2424      ENTER EMPLOYEE ID, THEN PRESS ENTER

        EMPLOYEE ID    2424

        EMP LAST NAME  TWO

        EMP FIRST NAME SAMPLE

        EMP SOCIAL SEC 747474747

        EMP YEARS SRVC 13

        EMP LAST PROM  2016-01-01




    MAKE CHANGES AND THEN PRESS PF5
    F2 INQ   F3 EXIT   F4 ADD   F5 CHG   F6 DEL

    EMPMDEL                    EMPLOYEE DELETE                        EMDE


        EMPLOYEE ->    2424      ENTER EMPLOYEE ID, THEN PRESS ENTER

        EMPLOYEE ID    2424

        EMP LAST NAME  TWO

        EMP FIRST NAME SAMPLE

        EMP SOCIAL SEC 747474747

        EMP YEARS SRVC 13

        EMP LAST PROM  2016-01-01




    PRESS PF6 TO DELETE THIS RECORD
    F2 INQ   F3 EXIT   F4 ADD   F5 CHG   F6 DEL
```

EMPPGDEL

Finally, let's test the delete screen. We'll delete transfer to the three other programs. We can use employee 3333.

```
EMPMDEL                    EMPLOYEE DELETE                           EMDE


        EMPLOYEE ->    3333      ENTER EMPLOYEE ID, THEN PRESS ENTER

        EMPLOYEE ID    3333

        EMP LAST NAME  RADISSON

        EMP FIRST NAME BENTLEY

        EMP SOCIAL SEC 777777777

        EMP YEARS SRVC 46

        EMP LAST PROM  2015-07-01




    PRESS PF6 TO DELETE THIS RECORD
    F2 INQ   F3 EXIT   F4 ADD   F5 CHG   F6 DEL
```

Now transfer to the inquiry screen with PF2.

```
EMPMINQ                    EMPLOYEE INQUIRY                          EMIN


        EMPLOYEE ->    3333      ENTER EMPLOYEE ID, THEN PRESS ENTER

        EMPLOYEE ID    3333

        EMP LAST NAME  RADISSON

        EMP FIRST NAME BENTLEY

        EMP SOCIAL SEC 777777777

        EMP YEARS SRVC 46

        EMP LAST PROM  2015-07-01




    F2 INQ   F3 EXIT   F4 ADD   F5 CHG   F6 DEL
```

Now transfer back to the delete program

```
EMPMDEL                      EMPLOYEE DELETE                            EMDE


        EMPLOYEE ->    3333       ENTER EMPLOYEE ID, THEN PRESS ENTER

        EMPLOYEE ID    3333

        EMP LAST NAME  RADISSON

        EMP FIRST NAME BENTLEY

        EMP SOCIAL SEC 777777777

        EMP YEARS SRVC 46

        EMP LAST PROM  2015-07-01




    PRESS PF6 TO DELETE THIS RECORD
    F2 INQ   F3 EXIT   F4 ADD   F5 CHG   F6 DEL
```

And then transfer to the change program using PF5.

```
EMPMCHG                      EMPLOYEE CHANGE                            EMCH


        EMPLOYEE ->    3333       ENTER EMPLOYEE ID, THEN PRESS ENTER

        EMPLOYEE ID    3333

        EMP LAST NAME  RADISSON

        EMP FIRST NAME BENTLEY

        EMP SOCIAL SEC 777777777

        EMP YEARS SRVC 46

        EMP LAST PROM  2015-07-01




    MAKE CHANGES AND THEN PRESS PF5
    F2 INQ   F3 EXIT   F4 ADD   F5 CHG   F6 DEL
```

Now transfer back to the delete program

```
EMPMDEL                    EMPLOYEE DELETE                          EMDE


        EMPLOYEE ->    3333       ENTER EMPLOYEE ID, THEN PRESS ENTER

        EMPLOYEE ID    3333

        EMP LAST NAME  RADISSON

        EMP FIRST NAME BENTLEY

        EMP SOCIAL SEC 777777777

        EMP YEARS SRVC 46

        EMP LAST PROM  2015-07-01




PRESS PF6 TO DELETE THIS RECORD
F2 INQ   F3 EXIT   F4 ADD   F5 CHG   F6 DEL
```

And then transfer to the add program. Notice we do not pass the employee number to the add program since there is no need to add an already existing employee.

```
EMPMADD                    EMPLOYEE ADD                             EMAD

                    ENTER EMPLOYEE INFO, THEN PRESS PF4

        EMPLOYEE ID

        EMP LAST NAME

        EMP FIRST NAME

        EMP SOCIAL SEC

        EMP YEARS SRVC

        EMP LAST PROM




ENTER DATA FOR NEW EMPLOYEE, THEN PRESS PF4 TO ADD
F2 INQ   F3 EXIT   F4 ADD   F5 CHG   F6 DEL
```

That's it, looks like everything works. Our integration test is finished.

Wrap-up

Well we've come to the end of this book. There are many other things you could do with this CICS project and many other things you can do with CICS. I left you with a fairly simple change to make to allow the programs to switch back to the main menu. I recommend that you perform that change.

I encourage you to use your imagination to extend this employee project. For example, you could create a EMP_PAY table and develop a payroll cycle with it. Then create CICS screens to review an employee's pay history. That's a meaty project.

Here are just a few more ideas you can consider. How about setting up archive tables for the EMPLOYEE and EMP_PAY tables – the archive records will come in handy when someone asked when these tables changed and what the previous entries were. Again, you can create CICS screens to review the history entries online. The possibilities are endless but the necessities will depend on your actual project requirements.

Let me close with a sincere "best of luck". It's been a pleasure walking you through CICS concepts with DB2 and VSAM. I truly hope you do exceptionally well as a CICS developer, and that you have every success!

Robert Wingate

IBM Certified Application Developer – DB2 11 for z/OS

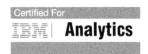

Test Yourself

Check yourself out on these questions and answers about CICS. First, here are the questions without answers.

Questions

1. What does the CICS acronym stand for?

2. In which programming languages can you develop CICS programs?

3. What is BMS?

4. In a COBOL program, how are CICS commands enclosed?

5. What are the CICS commands for handling screen interactions?

6. What does the symbolic map include?

7. What is the difference between a physical BMS mapset and a BMS symbolic mapset?

8. What are the three macros used in building a mapset?

9. What attributes are used to make a field normal intensity on the screen and protected (so that the field cannot change)?

10. What attribute will cause the cursor to be placed on a field when the screen displays?

11. Which EIB field contains the length of the data that was passed to the program?

12. The CICS commands for processing records from VSAM files are:

13. What typically causes a MAPFAIL condition when processing a RECEIVE MAP command?

14. The communication or COMMAREA must be included in the Linkage section of the program. What is the required name for this field in the Linkage section?

15. What is the syntax of the RECEIVE MAP command?

16. When do you need to use the NEWCOPY keyword?

17. What is the syntax of the XCTL command?

18. On a SEND command, what happens if you don't specify MAPONLY or DATAONLY?

19. What is the syntax of the READ command when you intend to update the record later?

20. If an application READS a VSAM KSDS file with UPDATE, and decides not to update the record, what command can be issued to release exclusive control on the record?

21. What exceptional condition will be raised if you try to insert a record into a file and a record with the same key is already there?

Questions with Answers

1. What does the CICS acronym stand for?

    CICS basically stands for Customer Information Control System.

2. In which programming languages can you develop CICS programs?

    COBOL, Assembler, PLI, Java and C/C++

3. What is BMS?

    BMS is Basic Map Support. It allows you to code assembler level programs to define screens.

4. In a COBOL program, how are CICS commands enclosed?

    CICS commands are coded between the EXEC CICS and END-EXEC statements. For example:

    ```
    EXEC CICS
        <command>
    END-EXEC
    ```

5. What are the CICS commands for handling screen interactions?

    RECEIVE MAP **– Retrieves input from the terminal.**

    SEND MAP **– Sends information to the terminal.**

6. What does the symbolic map include?

    It includes (in COBOL) two 01 level structures, one for input and one for output.

7. What is the difference between a physical BMS mapset and a BMS symbolic mapset?

The physical mapset is a load module used to map the data to the screen at execution time. The symbolic map is the actual copybook member used in the program to reference the input and output fields on the screen.

8. What are the three macros used in building a mapset?

    The three macros used in building a mapset are:

    - **DFHMSD starts the mapset.**

    - **DFHMDI starts a map within the mapset.**

    - **DFHMDF defines each field within a map.**

9. What attributes are used to make a field normal intensity on the screen and protected (so that the field cannot change)?

    The attribute value for this scenario is coded as:

    ```
    ATTRB=(NORM,PROT)
    ```

10. What attribute will cause the cursor to be placed on a field when the screen displays?

    Use the IC attribute value to cause the cursor to be placed on a field when the screen displays. For example:

    ```
    ATTRB=(NORM,UNPROT,IC)
    ```

11. Which EIB field contains the length of the data that was passed to the program?

    The EIBCALEN field contains the length of the data that was passed to the program.

12. The CICS commands for processing records from VSAM files are:

- **WRITE – Adds a record.**

- **READ – retrieves a record.**

- **DELETE – deletes a record.**

- **REWRITE – updates a record.**

13. What typically causes a MAPFAIL condition when processing a RECEIVE MAP command?

    When no data was sent from the screen, this raises a MAPFAIL **condition.**

14. The communication or COMMAREA must be included in the Linkage section of the program. What is the required name for this field in the Linkage section?

    The area must be defined as the first area in the Linkage Section and must be called DFHCOMMAREA.

15. What is the syntax of the RECEIVE MAP command?

    The syntax of the RECEIVE MAP **command is as follows:**

```
EXEC CICS
    RECEIVE MAP (map name)
                MAPSET(map set name)
                INTO (data name)

END-EXEC.
```

So for example if your mapset name is EMPMS01 and your map is named EMPM02, and your data structure name is EMPMAP1, you would code:

```
EXEC CICS
    RECEIVE MAP (EMPM02)
            MAPSET(EMPMS01)
            INTO (EMPMAP1)
END-EXEC.
```

16. When do you need to use the NEWCOPY keyword?

You use `NEWCOPY` **with** `CEMT` **to bring the latest version of the program from the loadlib into CICS.**

For example, to bring latest version of program `EMPPGM1` **into storage, issue:**

```
CEMT SET PROGRAM(EMPPGM1) NEWCOPY
```

17. What is the syntax of the XCTL command?

The syntax of the XCTL command is:

```
EXEC CICS
      XCTL PROGRAM (program name)
END-EXEC.
```

18. On a SEND command, what happens if you don't specify MAPONLY or DATAONLY?

Both constant data from the physical map and modifiable data from the symbolic map are sent.

19. What is the syntax of the READ command when you intend to update the record later?

The syntax of the READ command when you intend to update the record is as follows:

```
EXEC CICS
   READ FILE (file name)
   INTO (data structure name)
   RIDFLD(field name)
   UPDATE
   RESP(RESPONSE-CODE)
END-EXEC.
```

20. If an application READS a VSAM KSDS file with UPDATE, and decides not to update the record, what command can be issued to release exclusive control on the record?

Issuing an `EXEC CICS UNLOCK FILE` **(filename) command with the File or Dataset option will release control of the record. The lock will also be released if another** `READ` **is issued to move to another record.**

21. What exceptional condition will be raised if you try to insert a record into a file and a record with the same key is already there?

A `DUPREC` **condition will be raised if you try to insert a record into a file and a record with the same key is already there.**

Index

Additional Resources

IBM Introduction to CICS

https://www.ibm.com/support/knowledgecenter/zosbasics/com.ibm.zos.zmidtrmg/zmiddle_13.htm

CICS Transaction Server for z/OS® Application Programming Reference

https://www.ibm.com/support/knowledgecenter/en/SSGMCP_4.2.0/com.ibm.cics.ts.applicationprogramming.doc/topics/dfhp4_overview.html

IBM. (2017). *www.ibm.com*. Retrieved from DB2 11 for z/OS Documentation: https://www.ibm.com/support/knowledgecenter/en/SSEPEK_11.0.0/home/src/tpc/db2z_11_prodhome.html

IBM What is VSAM?

https://www.ibm.com/support/knowledgecenter/zosbasics/com.ibm.zos.zconcepts/zconcepts_169.htm

Other Titles by Robert Wingate

IMS Basic Training for Application Developers

ISBN-13: 978-1793440433

This book will teach you the basic information and skills you need to develop applications with IMS on IBM mainframe computers running z/OS. The instruction, examples and sample programs in this book are a fast track to becoming productive as quickly as possible using IMS with COBOL and PLI. The content is easy to read and digest, well organized and focused on honing real job skills.

Quick Start Training for IBM z/OS Application Developers, Volume 1

ISBN-13: 978-1986039840

This book will teach you the basic information and skills you need to develop applications on IBM mainframes running z/OS. The instruction, examples and sample programs in this book are a fast track to becoming productive as quickly as possible in JCL, MVS Utilities, COBOL, PLI and DB2. The content is easy to read and digest, well organized and focused on honing real job skills. IBM z/OS Quick Start Training for Application Developers is a key step in the direction of mastering IBM application development so you'll be ready to join a technical team.

Quick Start Training for IBM z/OS Application Developers, Volume 2

ISBN-13: 978-1717284594

This book will teach you the basic information and skills you need to develop applications on IBM mainframes running z/OS. The instruction, examples and sample programs in this book are a fast track to becoming productive as quickly as possible in VSAM, IMS and DB2. The content is easy to read and digest, well organized and focused on honing real job skills. IBM z/OS Quick Start Training for Application Developers is a key step in the direction of mastering IBM application development so you'll be ready to join a technical team.

DB2 Exam C2090-313 Preparation Guide

ISBN 13: 978-1548463052

This book will help you pass IBM Exam C2090-313 and become an IBM Certified Application Developer - DB2 11 for z/OS. The instruction, examples and questions/answers in the book offer you a significant advantage by helping you to gauge your readiness for the exam, to better understand the objectives being tested, and to get a broad exposure to the DB2 11 knowledge you'll be tested on.

DB2 Exam C2090-320 Preparation Guide

ISBN 13: 978-1544852096

This book will help you pass IBM Exam C2090-320 and become an IBM Certified Database Associate - DB2 11 Fundamentals for z/OS. The instruction, examples and questions/answers in the book offer you a significant advantage by helping you to gauge your readiness for the exam, to better understand the objectives being tested, and to get a broad exposure to the DB2 11 knowledge you'll be tested on. The book is also a fine introduction to DB2 for z/OS!

DB2 Exam C2090-313 Practice Questions

ISBN 13: 978-1534992467

This book will help you pass IBM Exam C2090-313 and become an IBM Certified Application Developer - DB2 11 for z/OS. The 180 questions and answers in the book (three full practice exams) offer you a significant advantage by helping you to gauge your readiness for the exam, to better understand the objectives being tested, and to get a broad exposure to the DB2 11 knowledge you'll be tested on.

DB2 Exam C2090-615 Practice Questions

ISBN 13: 978-1535028349

This book will help you pass IBM Exam C2090-615 and become an IBM Certified Database Associate (DB2 10.5 for Linux, UNIX and Windows). The questions and answers in the book offer you a significant advantage by helping you to gauge your readiness for the exam, to better understand the objectives being tested, and to get a broad exposure to the knowledge you'll be tested on.

DB2 10.1 Exam 610 Practice Questions

ISBN 13: 978-1-300-07991-0

This book will help you pass IBM Exam 610 and become an IBM Certified Database Associate. The questions and answers in the book offer you a significant advantage by helping you to gauge your readiness for the exam, to better understand the objectives being tested, and to get a broad exposure to the knowledge you'll be tested on.

DB2 10.1 Exam 611 Practice Questions

ISBN 13: 978-1-300-08321-4

This book will help you pass IBM Exam 611 and become an IBM Certified Database Administrator. The questions and answers in the book offer you a significant advantage by helping you to gauge your readiness for the exam, better understand the objectives being tested, and get a broad exposure to the knowledge you'll be tested on.

DB2 9 Exam 730 Practice Questions: Second Edition

ISBN-13: 978-1463798833

This book will help you pass IBM Exam 730 and become an IBM Certified Database Associate. The questions and answers in the book offer you a significant advantage by helping you to gauge your readiness for the exam, to better understand the objectives being tested, and to get a broad exposure to the knowledge you'll be tested on.

DB2 9 Certification Questions for Exams 730 and 731: Second Edition

ISBN-13: 978-1466219755

This book is targeted for IBM Certified Database Administrator candidates for DB2 9 for Windows, Linux and UNIX. It includes approximately 400 practice questions and answers for IBM Exams 730 and 731 (6 complete practice exams).

About the Author

Robert Wingate is a computer services professional with over 30 years of IBM mainframe programming experience. He holds several IBM certifications, including IBM Certified Application Developer - DB2 11 for z/OS, and IBM Certified Database Administrator for LUW. He lives in Fort Worth, Texas.

www.ingramcontent.com/pod-product-compliance
Lightning Source LLC
LaVergne TN
LVHW082033050326
832904LV00006B/271